ARISTOTLE'S LADDER,
DARWIN'S TREE

Aristotle's Ladder, Darwin's Tree

The Evolution of Visual Metaphors for Biological Order

J. David Archibald

COLUMBIA UNIVERSITY PRESS   NEW YORK

Columbia University Press
Publishers Since 1893
New York   Chichester, West Sussex
cup.columbia.edu

Library of Congress Cataloging-in-Publication Data
Archibald, J. David.
   Aristotle's ladder, Darwin's tree : the evolution of visual metaphors
for biological order / J. David Archibald.
        pages cm
    Includes bibliographical references and index.
    ISBN 978-0-231-16412-2 (cloth)
    ISBN 978-0-231-53766-7 (e-book)
  1. Biology—Philosophy. 2. Human evolution—Philosophy. 3. Imagery (Psychology)
4. Metaphor. I. Title.
   QH315.A723 2014
   570.1—dc23

                                                            2013050622

COVER IMAGE: Frontispiece of Anna Maria Redfield, *Zoölogical Science, or, Nature in Living Forms*
(New York: Kellogg, 1858). (Courtesy of the author)

COVER AND TEXT DESIGN: Milenda Nan Ok Lee

For my father,

James R. Archibald (1927–2013)

Contents

Preface

Every culture that has put chisel to stone or pen to paper has attempted to visualize the order in nature and our place in it. Some of the more intriguing representations of the natural biological order remain with us in the form of grandiose spirals hypothesizing the relationships of thousands upon thousands of species. We have come a long way, but we've become so comfortable with these representations that we must remind ourselves that they are poetic metaphors rather than the scientific history of life on planet Earth. We blithely presume an underlying reality—that this natural biological order came about by the process of evolution. This realization emerged succinctly in only the past two centuries, whereas our graphic images and schemata remain only metaphors for this process and for the pattern or patterns that emerged over the past few thousand millennia. Shockingly, even in the self-described advanced cultures the very fact of evolution remains controversial mostly because of religious zealotry and ignorance.

The acceptance of evolution as the greatest force underlying nature emerged in Europe as science ascended the remaining steps to the throne of rationality. As might be expected, the ideas of evolution and in particular its visualization did not suddenly appear treading on the heels of the Enlightenment; rather, they exhibit a long and sometimes tangled history within the Western tradition. Ladders and trees became the common but not the only icons. The growth and blossoming of visual representations over the past 2,500 years and what they meant to those who created them encompass the theme of this book.

How we visualized nature's order leading up to modern evolutionary biology by necessity includes or excludes various individuals and their ideas in a somewhat idiosyncratic manner. This pertains especially to important biologists from the nineteenth century onward. Accordingly, it must be emphasized that the basis for including or excluding an individual relates most specifically to whether this individual contributed a visual representation of biological order or in a few cases a written description of how to represent biological order.

Placing past events within the context of time and place proved a daunting task. Supposedly the more we know, the easier the task becomes. But we then face the situation of more experts weighing in on the meaning of this visual metaphor or that narrative. It is, to be sure, a dubious, untidy process for us humans to try to objectify ourselves, because we deem ourselves exceptions to the rest of the natural world. Having spent most of my career studying long-dead species that never possessed such an exalted view (if they possessed any view) of themselves, I fortunately could eliminate the issue of hubris in the subjects of study.

Studying long-dead creatures and their environments presents many variables, but one is not their view of their place in nature. Especially given the newer techniques of placing long-extinct species within their environmental context, with the overlay of knowing that all such species remain subject to the ravages and rewards of evolution, we paleontologists pride ourselves in placing organisms in a proper context of time and place. Such is not the case when interpreting human history, even when written. We run the risk of trying to place these sometimes very ancient ideas within in a modern context and placing modern sensibilities on them.

In this book, then, the task is to see a diagram not as we now interpret it but as the author and intellectually curious consumers at the time perceived it. Such diagrams—ladders, stairs, trees, tables, bifurcating figures—meant one thing at the time they were created but, depending on the longevity of the figure, have affected both how we draw such diagrams today and how we interpret them. Does a tree figure with various species at its termini mean an evolutionary history? Does a simple bifurcating diagram of various species represent a genealogy? When we see diagrams of fish to amphibian to reptile to mammal to human, what do we perceive these representations mean? The answer depends on when and where the diagram appeared.

How do we measure progress in our understanding of the biological order? We can identify benchmarks—among others, Lamarck's use-disuse ideas and his "tree" in 1809; Darwin's natural selection and his hypothetical tree in 1859; Haeckel's many phylogenies in the latter nineteenth century; Simpson's help in reconciling Mendel and Darwin in the mid-twentieth century; and the rapid-fire introduction toward the end of the twentieth century of Hennig's cladistics, PCR, and related molecular techniques for phylogenetic reconstruction, and the repeatedly doubling of computing power. Building a scale of progress based on this sort of trajectory, we made slow progress for more than two thousand years, began picking up speed just over two hundred years ago, and turned vertically about fifty years ago. As we shall see, the visual metaphors for this incredible progress have not kept pace, and understandably so. As the number of species that we believe to exist and those we think exist took an equally astounding upward turn, our ability to put this in visual metaphors flagged. Who can grasp millions of interrelated species festooning the tree of life? Our computers can calculate and even attempt to draw such trees, but we must at some point simply look on in awe.

Acknowledgments

I thank Patrick Fitzgerald, my editor at Columbia University Press, for his early and strong support for this project, as well as two anonymous reviewers for their thoughtful comments. Bridget Flannery-McCoy, associate editor for the sciences, and Milenda Lee, senior designer at Columbia, offered their wonderfully helpful advice and thoughts in the preparation of this volume, as did Columbia editors Kathryn Schell and Irene Pavitt. Anita O'Brien provided expert copyediting skills. Curtis Johnson gave useful input on chapter 4. Gloria E. Bader and E. N. Genovese read the manuscript in great detail, supplying excellent critiques on language, syntax, and content, and provided suggestions for titles. For this I am especially grateful. Edward Cell read and commented on the completed text.

A wealth of information came from other people. Giulia Caneva freely shared with me her ideas and interpretations of the vegetal friezes on the Ara Pacis. David McLoughlin inspired me to relate the apse mosaic in San Clemente to the Ara Pacis vegetal frieze. Sara Magister similarly directed me to the vegetal-motif mosaic occupying the apse in the narthex, or antechamber, to the Baptistery of the Papal Archbasilica of St. John Lateran. Luca Dejaco kindly provided a copy of the *Bell'Italia* article on the floor mosaic of the Cathedral of Otranto. John van Wyhe introduced me to the work of Anna Maria Redfield. Thanks to Malcolm Kotter for alerting me to the evolutionary trees in George William Hunter's *A Civic Biology*.

A number of online sources proved immensely helpful: the American Museum of Natural History Darwin Manuscripts Project, the Biodiversity Heritage Library, the Darwin Correspondence Project, Darwin Online, and Google Books. I greatly appreciate the American Philosophical Society Library, the Cambridge University Library, the Huntington Library, and their staffs for allowing me to use materials in their collections. I thank San Diego State University Interlibrary Loan for obtaining copies or originals of materials used in this book.

Acknowledgments for the use of images are provided in the captions of respective figures. Other images not acknowledged either are not under copyright or are photographs by the author. Images not under copyright come from a private library, from Biodiversity Heritage Library, from Google Books, or from Wikimedia Commons.

ARISTOTLE'S LADDER,
DARWIN'S TREE

CHAPTER ONE

Blaming Aristotle

Our perceptions as well as our misperceptions of the history of life on this planet arise in large measure from the representations of evolutionary history, both verbal and visual. One need not be a biologist to understand the meaning of "lower" and "higher" animals. Images abound showing the march of primate evolution from a lowly, monkey-like ancestor to the pinnacle of humanness—*Homo sapiens*. We do, of course, deem ourselves as the highest animals—in the Western tradition, just below the angels. But what do we mean with these seemingly innocuous adjectives? What makes us presume that we are the highest of animals: are we closer to God, are we more complex, are we more highly evolved?

Natura Non Facit Saltus

We can blame Aristotle (384–322 B.C.E.). Aristotle's views come to us in his ten books titled *Researches About Animals*, more commonly known from the Latin translation *Historia Animalium* (*The History of Animals*). His classification of life accorded with the then accepted views of the four basic elements of nature (air, fire, water, earth). Aristotle defined groups often in apposition, such as "bloodless" animals and "blooded" animals, which basically correlate today with what we call invertebrates and vertebrates. These two groups were then further subdivided into what he called observable "forms" (*eidê*), larger groups, or "kinds" (*genê*). The Latin words *species* and *genera* only loosely correspond to what we today mean by species and genera. Although the relative hierarchy of species as subsets of genera still pertains, Aristotle used these terms for much larger sets of animals (Mayr 1985; Gagarin 2009).

Of particular interest here, Aristotle also provided the first surviving attempts in the Western world to arrange inanimate and animate objects in some ordered sense based on their level of complexity. Even if images of his system ever existed, none survives, yet his brief description suffices to make it quite clear what he intended:

> Nature proceeds little by little from things lifeless to animal life in such a way that it is impossible to determine the exact line of demarcation, nor on which side thereof an intermediate form should lie. Thus, next after lifeless things in the upward scale comes the plant, and of plants one will differ from another as to its amount of apparent vitality; and, in a word, the whole genus of plants, whilst it is devoid of life as compared with an animal, is endowed with life as compared with other corporeal entities. Indeed, as we just remarked, there is observed in plants a continuous scale of ascent towards the animal. (Aristotle 2007:8.1)

Certainly species across groups share characters; thus Aristotle sees the scale or ladder as forming a continuum, a succession without gaps from inanimate objects through plants and then to animals, thus *natura non facit saltus* (nature makes no leaps). Boundaries between groups do occur; we simply cannot discern them because of the continuous nature of characters shared by the various groups (Balme, in Aristotle 1991). Aristotle provides us with an explicit statement concerning the *scala naturae*, but he does not propose an evolutionary basis for this continuity. Aristotle did not support claims of earlier ideas of evolution made by other Greeks; such claims using Aristotelian *scala naturae* come much later. Rather, for Aristotle all was cyclic, with no beginning and no end.

Aristotle greatly influenced later writers on the same topic. Some four hundred years later, the Roman Pliny the Elder (Gaius Plinius Secundus, 23–79 c.e.) organized his thirty-seven-volume *Naturalis Historia* (*Natural History*) along the lines of Aristotlean *scala naturae*. Although the name would imply a work concerned with what now we call natural history, Pliny produced a far broader work that included various aspects of Roman culture. We do not know if Aristiotle would have approved, but unfortunately, as with Aristotle, if Pliny produced any stairs of nature they do not survive.

Pliny shared his Stoic philosophy (that misfortune and virtue are sufficient for happiness) with the Roman consul and orator Marcus Tullius Cicero (106–43 b.c.e.), specifically that purpose and design exist in nature, including humans' place within it. These views carried forward as Christianity came to political and social power in Europe. It must be said that these views, while monolithic, were not universal. Not all Romans of similar antiquity shared Pliny's Stoic approach. The first-century b.c.e. Epicurean Roman philosopher Lucretius (Titus Lucretius Carus, ca. 99–ca. 55 b.c.e.) accepted that there are gods but that they have no interest in humans, that the universe has no creator and was not created for humans, and that nature ceaselessly experiments (Greenblatt 2011). Troublemakers always nip at the heels of authority. Nevertheless, the Stoic philosophies and Aristotelian *scala naturae* held sway for the next millennium and a half but not in the way Aristotle first articulated it.

Extending the Ladder to Heaven

In late Christian Rome and into the Middle Ages, Aristotle's scale (ladder, stairs) expanded beyond the earthly realm into heavenly matters, with some rather interesting results. In this incarnation, the phrase the "great chain of being" is often applied (Lovejoy 1942). Paul Carus (1900) presents us with a rather comical extreme that he titles "Satanic Temptations and the Ladder of Life," which comes from the encyclopedic illuminated manuscript *Hortus deliciarum* (*Garden of Delights*, ca. 1161–1185) by Herrad von Landsberg (1130–1195), a twelfth-century abbess or mother superior at the Hohenburg (now Mont Sainte-Odile) Abbey in French Alsace. The book was written for the edification of novices. The version of the ladder shown in figure 1.1 includes descriptions in Latin of what transpires in the figure (Green et al. 1979). Her ladder includes only the portion of the great chain of being that deals with human frailties and the steps to heaven, so strictly speaking it is not an Aristotelian ladder. The text indicates rungs to heaven: purity, contempt of the world, humility, obedience, patience, faith, and love of a pure heart. In the upper left of the figure angels battle demons to protect the novices from the temptations of city life, precious garments, money, the couch of laziness, the joy of gardening, and possibly worst of all, the allure of worldly comforts that militaristic abbots might wield against an unwary novice (Carus 1900).

Like Herrad von Landsberg, the sixteenth-century Franciscan friar Diego Valadés (1533–1582) wrote a large tome for his fellow brethren, but unlike Herrad's *Hortus deliciarum*, which was to help novices, Valadés's *Rhetorica Christiana ad concionandi et orandi usum* (1579) addressed missionaries on how to educate and convert Native Americans. Valadés was the son of a Tlaxcaltecan Indian woman and a Spaniard who had arrived in Mexico with Cortez. Valadés became a Franciscan friar, spending time as a missionary in his native Mexico before going to Rome, where he wrote the *Rhetorica* with the intention not only of helping missionaries convert Native Americans but also of helping these missionaries better understand Native American peoples (Alejos-Grau 1994; Fane 1997). For example, three mundane figures show the Latin alphabet, using various familiar objects to illustrate the letters. As far as is known, Valadés created all the illustrations used in his volume.

The most often reproduced figure from *Rhetorica* is more ethereal (figure 1.2); it clearly shows the great chain of being, although often identified as representing creation (Alejos-Grau 1994). A scroll surrounding the top of the figure reads (in faulty Latin), "Ego sum principium et finis et preter me non est deus que [*sic*] omnes dii gentium demonia" (I am the beginning and the end and beyond me there is no god, and all the gods of the nations are demons). Just below the scroll is the Trinity—God on his throne, the Holy Spirit as a dove, and Jesus Christ, with Mary on their right. Surrounding these figures, six archangels waft burning incense toward the Holy Family. A chain descends from God's right hand,

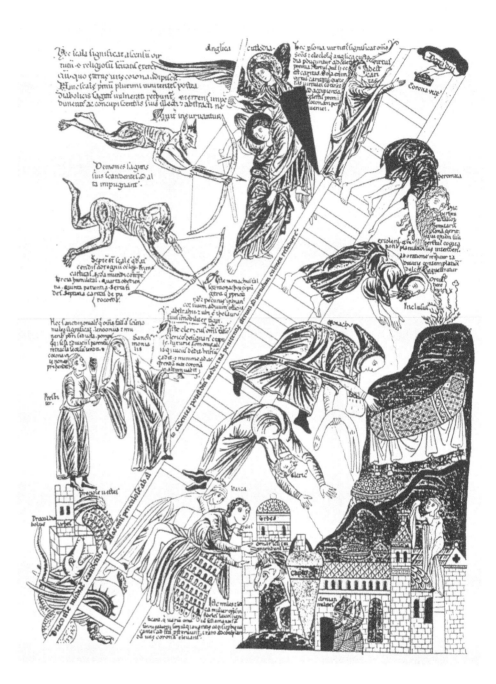

FIGURE 1.1 "Satanic Temptations
and the Ladder of Life" (so named
by Carus 1899), from Herrad von
Landsberg's *Hortus deliciarum*
(ca. 1161–1185).

figuratively if not literally anchoring the image as it passes downward through the
center of the illustration, terminating in hell at the head of Satan. Surrounding
Satan, various souls undergo forms of torture. Immediately above hell, an anchor-
like scroll festoons the image, across which is written "F Didicus Valadés fecit
(Friar Diego Valadés made this). In a rather tree-like fashion, arms protrude from
either side of the chain.

The base of the drawing above hell shows five circles representing the creation
of the stars, the separation between land and water, and that between chaos and

FIGURE 1.2 Diego Valadés's great chain of being, from *Rhetorica Christiana ad concionandi et orandi usum* (1579).

the world. The second from the right holds a bishop, possibly Saint Augustine, and a child trying to understand the mystery of the Trinity (Alejos-Grau 1994). Above this, the creation of life-forms commences with various plants, followed next upward by two side arms bearing a menagerie of real land mammals (bear, deer, goat, camel, elephant) mingled with mythic animals (unicorn, dragon). Sea creatures—including bivalves, fish, turtles, and a whale—populate the next higher arms. Above this, the level shows various birds walking and flitting about. Interestingly, Valadés illustrates animals and plants from the five known continents, including for the first time some indigenous American animals and plants, such as the quetzal, turkey, llama, cactus, maize, cocoa, and pineapple. Above the birds, we finally reach humans, whose dress indicates peoples from around the world, including Native America. Valadés's intent shows the Americas as being as old as the rest of the world and the Native Americans as part of this unity and universality of humankind, as descendants of Adam and Eve, who appear in the middle of this level with Eve arising from Adam's side (Alejos-Grau 1994). At the penultimate level, below God, we find haloed and winged angels. We cannot forget the poor fallen angels being cast from heaven on the right, who increasingly appear less angelic as they fall, noticeably replacing their angel wings with bat-like and then fairy-like wings.

The image intends to show an upward progression from plants; through animals, humans, and angels; and then finally to heaven. The levels progress, but do not evolve, toward perfection. One problem: mammals rest lower than the fishes, a very un-Aristotelian portrayal of the ladder of life. Also, the tree-like appearance of the image is possibly illusory, if for no other reason than Valadés has two other, more tree-like figures in *Rhetorica*, the first titled the "Ecclesiastical Temporal Hierarchy" (figure 1.3*A*) and the second the "Ecclesiastical Hierarchy" (figure 1.3*B*).

FIGURE 1.3 Valadés's (*A*) "Ecclesiastical Temporal Hierarchy" and (*B*) "Ecclesiastical Hierarchy," from *Rhetorica Christiana ad concionandi et orandi usum* (1579).

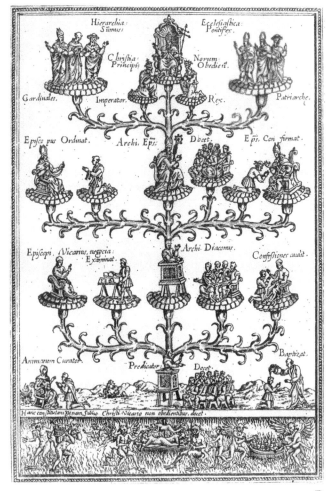

A

B

Notably, both images are subtended by the image of hell, as in Valadés's great chain of being or creation story, certainly placed there as warning, but again not Aristotelian. The meaning of each image is well described by its title. The temporal tree is surmounted by the emperor flanked by kings on each side, with lesser leaders on lower branches. The ecclesiastical hierarchy has at its head the pope, labeled "Pontifex," the Latin title for the pre-Christian chief religious official in ancient Rome. At his feet on bended knee are the emperor and the king, clearly placing the temporal hierarchy as subservient to the church. On lower branches are arrayed various other church functionaries. Valadés almost certainly chose the chain of being for his representation of all creation because that was the accepted view at the time. The reason for the use of more tree-like structures for the two hierarchies is less clear. One cannot discount the possibility of simple artistic license, because recall that Valadés was the illustrator of *Rhetorica*. As Carmen Alejos-Grau (1994) notes, the arrival of the Spaniards in the Americas portended that the natives should undergo a completely new type of a civilian government and new Christian faith. These trees, as she refers to them, were to this end.

A Return to Aristotle, but with a Twist

Although one of the better known, Valadés's *Rhetorica Christiana* (1579) was neither the first nor the last religious work to use the imagery of the great chain of being. Some 166 years later, in 1745, the Swiss philosopher and naturalist Charles Bonnet (1720–1793) created a great chain of being for the natural world in the first volume of his two-volume work on insects, *Traité d'insectologie* (*Treatise on Insects*, 1745) (Anderson 1976). The end of the preface of the first volume includes a large, narrow, foldout diagram shown in figure 1.4 in two parts, with the upper part on the right. Bonnet titles his diagram "Idée d'une échelle des êtres naturels" (Idea of a Scale of Natural Beings). Loosely translated from the French, he writes:

> This reflection has made me think, perhaps foolhardily, to draw up a ladder of natural beings, that we find at the end of this Preface. I produce it only as a trial, but suitable for conceiving of the ideas of the system of the World & the Infinite Wisdom which has formed & combined the different parts. Let us pay attention to this beautiful splendor. Let us look at the innumerable multitude of organized and unorganized bodies; to place one above the other, depending on the degree of perfection or excellence that is in each. If the sequence does not appear to us everywhere equally continuous; it is because our knowledge is still very confined: the more it increases, the more steps or degrees we will discover. . . . And if, as I think, all these scales, whose number is almost infinite, not only in form that combines all the possible orders of perfection, it must be admitted that one cannot conceive of anything greater or more exalted. . . . There is thus a connection between all parts of this universe. The system generally consists of the assembly of individual systems, which are like the different wheels of a machine. An insect, a plant is a particular system, a small wheel that in fact moves the greater. (xxviii–xxxi)

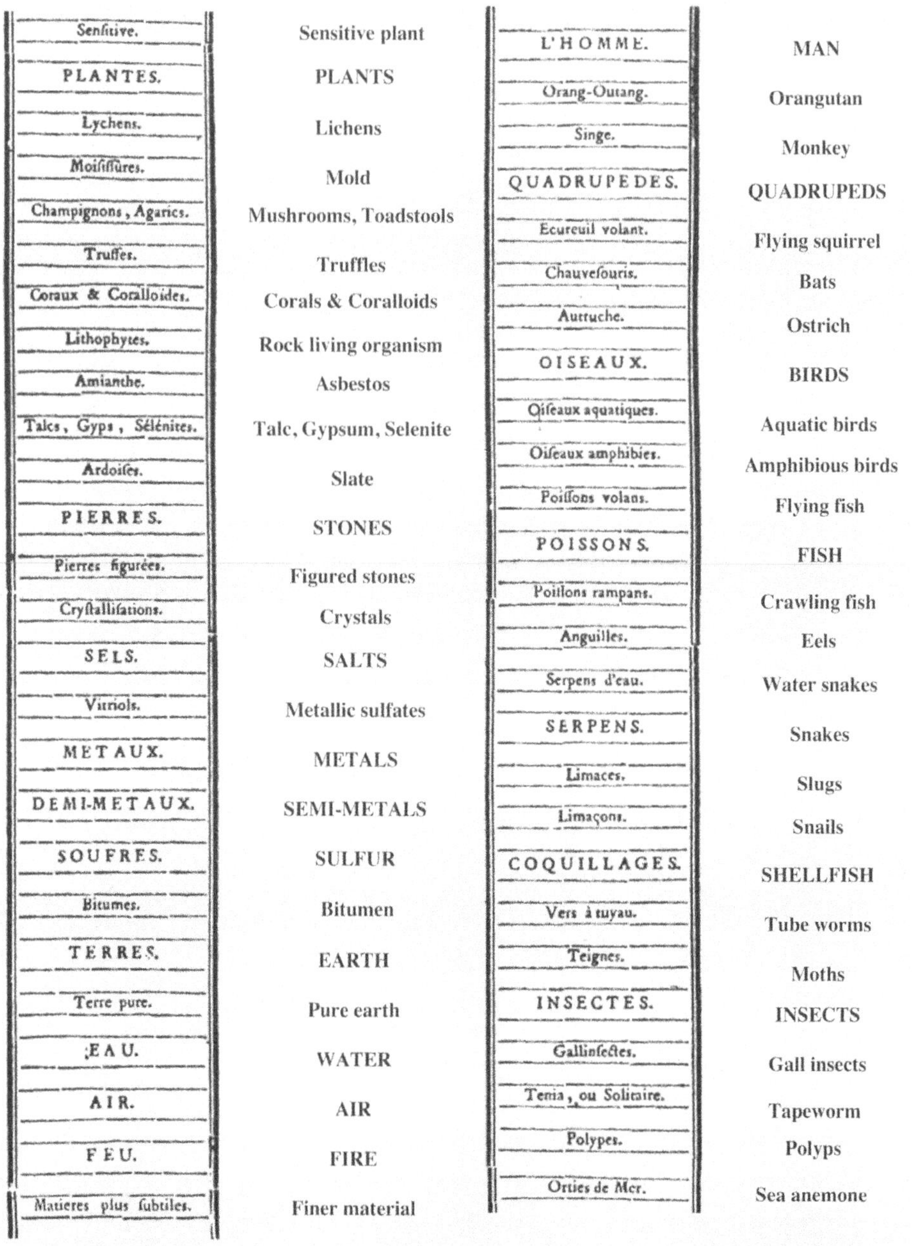

Senfitive.	Sensitive plant	L'HOMME.	MAN
PLANTES.	PLANTS	Orang-Outang.	Orangutan
Lychens.	Lichens	Singe.	Monkey
Moififfures.	Mold	QUADRUPEDES.	QUADRUPEDS
Champignons, Agarics.	Mushrooms, Toadstools	Ecureuil volant.	Flying squirrel
Truffes.	Truffles	Chauvefouris.	Bats
Coraux & Coralloides.	Corals & Coralloids	Autruche.	Ostrich
Lithophytes.	Rock living organism	OISEAUX.	BIRDS
Amianthe.	Asbestos	Oifeaux aquatiques.	Aquatic birds
Talcs, Gyps, Sélénites.	Talc, Gypsum, Selenite	Oifeaux amphibies.	Amphibious birds
Ardoifes.	Slate	Poiffons volans.	Flying fish
PIERRES.	STONES	POISSONS.	FISH
Pierres figurées.	Figured stones	Poiffons rampans.	Crawling fish
Cryftallifations.	Crystals	Anguilles.	Eels
SELS.	SALTS	Serpens d'eau.	Water snakes
Vitriols.	Metallic sulfates	SERPENS.	Snakes
METAUX.	METALS	Limaces.	Slugs
DEMI-METAUX.	SEMI-METALS	Limaçons.	Snails
SOUFRES.	SULFUR	COQUILLAGES.	SHELLFISH
Bitumes.	Bitumen	Vers à tuyau.	Tube worms
TERRES.	EARTH	Teignes.	Moths
Terre pure.	Pure earth	INSECTES.	INSECTS
EAU.	WATER	Gallinfectes.	Gall insects
AIR.	AIR	Tenia, ou Solitaire.	Tapeworm
FEU.	FIRE	Polypes.	Polyps
Matieres plus fubtiles.	Finer material	Orties de Mer.	Sea anemone

FIGURE 1.4 Charles Bonnet's "Idea of a Scale of Natural Beings," from *Traité d'insectologie* (1745). The left half is the lower part and the right half is the upper part of Bonnet's figure. (The English translations were added by the author.)

In a footnote to this passage, Bonnet (1745) writes, "If the greatest poets of our century, a Pope, a Voltaire, a Racine, wished to practice on a worthy subject, and give us the Temple of Nature, I think their work not only might be extremely useful but generally pleasing" (xxix). Almost sixty years later, the English polymath and physician (and grandfather of Charles Darwin) Erasmus Darwin (1731–1802) accommodated Bonnet by writing his long poem *Temple of Nature*, published in 1803, a year following his death. In his works, Darwin elaborates on clearly evolutionary ideas, although the word "evolution" was not used. Arguably, Bonnet did not hold Darwin's view of the *Temple of Nature* in which more complex life arose from simpler life.

It can be argued, however, that Bonnet accepted a kind of evolution but meant it in its original scientific sense of unrolling, a preformationist idea in which each succeeding generation was preformed in the sperm or egg and simply grew or unrolled (evolved). The similar idea of Bonnet's termed *emboîtement*, meaning "interlocking" or "nesting," argues that the germ cells of one generation contained the germ cells from which all successive generations arose. Bonnet's theory posited that the earth was episodically racked by catastrophes, the Noachian flood being just the latest. Although creatures died in these catastrophes, their germ lines survived to be part of a new creation. With each catastrophe, the newly created forms were higher in the scale of natural beings. This idea of a catastrophic end followed by a resurrection of sorts was a secularized version of clearly religious ideas. For Bonnet, the continuity of steps in his scale of natural beings formed a rational basis for his ideas of progressive change (O. W. Holmes 2006).

Bonnet's scale of natural beings accords well with the Aristotelian idea of a scale of nature without referring to religion. At the bottom of Bonnet's diagram, the four basic elements of nature noted by the ancient Greeks appear—fire, air, water, and earth (see figure 1.4, *left column*). Next upward follows what Bonnet perceived as less well-organized, nonliving substances such as sulfur, and then better organized, nonliving substances such as crystals. Probably because the skeletons of corals are rock and are sessile, Bonnet regards them as a transition from nonliving to living nature. These are followed upward by various fungi that in the eighteenth century qualified as plants, and then come what we today recognize as plants. When Bonnet reaches animals (*right column*), some fanciful transitions and juxtapositions occur moving upward. Shellfish are followed by snails, which are followed by slugs. In this sequence, shells disappear as bodies elongate. Continuing the trends, but leaving some levels out, we transition through snakes, water snakes, crawling fish, fish, flying fish, aquatic birds, birds, and so on. With what we would regard as tortured logic, he traverses ever upward through flying fish, various bird grades, bats, and flying (read "gliding") squirrels, and takes us back to terra firma with quadrupedal mammals. Even though illogical to us, the only true oddity in this part of the sequence is the flightless ostrich. Of course, man tops the ladder.

Bonnet creates a finely graded level of organization, although today it appears quite peculiar to us. Nowhere in Bonnet do we see the ladder stretching ever upward through temporal and ecclesiastical hierarchies as in Valadés's chain of being. Bonnet in no manner rejected such exalted hierarchies; rather he strived to examine the natural world much as his European contemporaries and arguably as Aristotle had done over two thousand years earlier. Bonnet has, however, also added his views of evolutionary change to explain Aristotle's ladder—catastrophe and death, rebirth, and successive progression and perfection after each catastrophe. For better or worse, Aristotle's scale of nature took on new meaning.

The closest Bonnet comes to Valadés's great chain of being occurs in a much later stairway-like engraving, reportedly by the Danish engraver Johan Frederik Clemens (1749–1831), who did various vignettes for Bonnet (figure 1.5). Unlike Bonnet's earlier *scala naturae*, which Bonnet specifically describes and which occurs in all formats of his work, the stairway-like engraving occurs in only one work and

in only one format. It appears near the beginning of the fourth volume of the quarto format of his *Oeuvres d'histoire naturelle et de philosophie*: *Contemplation de la nature* (*Works of Natural History and of Philosophy*: *Contemplation of Nature*, 1781). The first edition of *Contemplation* (1764) lacks the stairway-like engraving, as does the smaller octavo edition of Bonnet's (1781) works. This suggests artistic license on the part of Clemens (or whomever the artist was) in that in the larger quarto format he found greater space to express visual ideas, as he did elsewhere in the same work.

Although Bonnet does not directly address this figure, the artist likely took inspiration from statements by Bonnet (1781), such as just below the figure: "I am looking for relationships which make this huge chain a single All: I stop to consider some of the links, and am struck by the power of these traits, and the wisdom and magnitude of what I discover. . . . To make the eternal Universe eternal, is to admit an infinite succession of finite beings" (1–2). These words are the same in the text of the first volume of *Contemplation* (1764) and in the smaller edition of Bonnet's (1781) works, but again the figure is not present in these other formats. In the figure, a man stands at the top of the stairway with his head in the clouds, certainly an allusion to his place in both the physical and spiritual worlds (see figure 1.5). Arrayed down the stairs are a monkey, a lion, a dog; then birds, fish, and insects; and finally plants and crystals. Everything is surrounded by clouds such that, intended or not, this engraving from 1781 invokes a more metaphysical aspect not seen in the ladder-like figure of 1745.

For Bonnet, the ladder, scale, or chain presented infinite, all-encompassing grades but known only imperfectly to human understanding (Hopwood, Schaffer, and Secord 2010). For Bonnet, the totality of nature forms a hierarchical continuum (Heilbron 1990). Bonnet provided visualization of his "scale of natural beings," but he was by no means the only naturalist in the seventeenth and early eighteenth centuries to contemplate such a scale or to toy with some form of evolution; Gottfried Wilhelm Leibniz (1646–1716), Georges-Louis Leclerc, Comte de Buffon (1707–1788), Jean-Baptiste Lamarck (1744–1829), Gottfried Wilhelm Leibniz (1646–1716), and Étienne Geoffroy Saint-Hilaire (1772–1844) represent some of the most prominent of such scientists. Yet none provided us with visualizations of the "scale of natural beings," although as we will see in chapter 3, Lamarck was the first to provide an evolutionarily based tree or what we now call a phylogeny.

Dismantling Ladders and Smashing the Glass Jars

The great French anatomist and paleontologist Georges Cuvier (1769–1832) would have none of it. He rejected the idea of the existence of a *scala naturae* possessing gaps, and he rejected that evolution had occurred; yet he could not be strictly called a creationist (Taquet 2006, 2009). The essence of these ideas can be found in a few short passages in *Memoirs of Baron Cuvier* (1833) by Mrs. R. Lee (Sarah Bowdich Lee, 1791–1856), published just one year after Cuvier's death. Lee translates long passages from Cuvier's (1825) article "Nature" in the *Dictionnaire des sciences naturelles*. Through her extended translations of parts of his work, we gain a clear picture of his antipathy to these ideas.

About Cuvier's "Nature" article, Lee (1833) notes that "it contains the clearest and most satisfactory refutation of the reigning controversies that has ever been published in a separate form" (139). Cuvier agrees that laws of motion presiding over organization exist but that "a great many writers . . . have suffer[ed] themselves to be drawn unconsciously towards doctrines which have no other foundation. Such are the doctrines of the 'Scale of Nature,' the 'Unity of composition,' and others similar to these, which have all been imagined in consequence of the belief in a Nature distinct from the Creator" (142–43). Later, Cuvier waxes even more strongly:

> that the forms of these beings necessarily constitute a series or a chain, so that the eye may gradually pass from one to the other, without finding any gap, any hiatus; in short, the existence of a continued and regular scale in the forms of beings, from the stone to the man . . . this is what is not true, whatever eloquence may have been used in tracing the imaginary picture. The philosophers who have supported this system of a scale of beings, at each interruption which is pointed out to them, pretend, that if a step is wanting, it is hidden in some corner of the globe, where a fortunate traveller may one day discover it. Nevertheless, all regions, all seas, have been explored; the number of species collected increases every day; there are, perhaps, a hundred-fold more than when these paradoxical opinions began to be established, and none of the spaces are filled up; all the interruptions remain; there is nothing intermediate between birds and other classes; there is nothing between vertebrated animals and those which have no vertebrae. (144–45)

Next Cuvier takes on the idea of evolution:

> Nevertheless, to the hypothesis of a continued scale in the forms of beings, other philosophers have added that in which all beings are modifications of one only; or, that they have been produced successively, and by the development, of one first germ; and it is on this that an identity of composition for all has been engrafted. . . . This system (as it now exists) seizes hold of some partial resemblances, without having any regard to differences; it sees in the worm the embryo of the vertebrated animal; in the vertebrated animal with cold blood, the embryo of the animal with warm blood; it thus makes one class spring from the other. . . . We, however, conceive nature to be simply a production of the Almighty, regulated by wisdom, the laws of which can only be discovered by observation; but we think that these laws can only relate to the preservation and harmony of the whole; but we do not perceive any necessity for a scale of beings, nor for a unity of composition, and we do not believe even in the possibility of a successive appearance of different forms; for it appears to us that, from the beginning, diversity has been necessary to that harmony, and that preservation, the only ends which our reason can perceive in the arrangement of the world. (147, 150–51)

It would seem that Cuvier rejected evolution, but in fact he possessed more nuanced views (Taquet 2006, 2009). Cuvier was born into a Lutheran family in what is now eastern France. Early in his life, he sat for exams for theological study at Tübingen but was not accepted, a fortunate failure because of the later impact he had on the nascent sciences of anatomy, paleontology, systematics, and geology. The death of all four of his children may well have colored his worldview, but he appears to have believed in supernatural design yet was familiar with critiques against design put forth by Kant, Buffon, and d'Holbach (Taquet 2009). Cuvier was clearly hostile to the materialism that he saw in evolutionary theorizing of the time, especially that of his one-time mentor Lamarck; but this does not mean he felt that species were directly created by a god but rather he remained quite neutral on proximate and ultimate causes regarding the origin of new species (Rudwick 1998).

Nonetheless, as the quotes from Lee (1833) demonstrate, Cuvier rejected the scale of being. But what did he propose as a substitute? Proceeding from his earlier anatomical studies, Cuvier deduced that animal life formed a series of four *embranchements* (branches) that he proposed in a short paper in 1812, but later expanded in his four-volume work colloquially known as *Le Règne animal* (*The Animal Kingdom*, 1817). In his 1812 paper he writes:

> Considering the animal kingdom in this new perspective, and having regard to animals themselves, not their size, their usefulness, in varying degrees of knowledge which we, nor the other incidental circumstances, I have found that there are four principal forms, four general plans, after which all animals appear to have been modeled, and whose subsequent divisions, a few names which naturalists have furnished, are only slight modifications based on the development or the addition of a few parts, but do not change the essence of the plan. (77)

The new perspective of which he writes concerns treating all vertebrates as one branch that is equivalent to three other branches of invertebrates, which were traditionally lumped together. In addition to Vertebrata, Cuvier recognized Articulata (segmented worms and arthropods), Mollusca (mollusks plus other soft, bilaterally symmetrical invertebrates), and Radiata (corals and relatives, and echinoderms). Only one of these branches, as Cuvier envisioned them, is today regarded as a natural, evolutionary group: the vertebrates. All the others include mixtures of unrelated forms. Cuvier interpreted each branch as having a fundamentally different body plan from the others.

Life could no longer be viewed as a solid existence without gaps or a ladder with infinite rungs; Cuvier had dismantled it. As dramatically expressed by Michel Foucault (1970), in an iconoclastic gesture Cuvier had smashed the metaphorical glass jars of gapless continuity portrayed in the gardens and museums of the time, replacing them with his four *embranchements* that no longer obscured anatomy and function of plants and animals (Hopwood, Schaffer, and Secord 2010). To my knowledge, Cuvier never produced a tree-like diagram showing these four branches, but his choice of the word *embranchement* strongly invokes an image of a branching form such as a tree. Later writers—such as the American geologist Edward Hitchcock (1793–1864), the Canadian naturalist Anna Maria Redfield (1800–1888, née Treadwell), and the Swiss American paleontologist and protégé of Cuvier, Louis Agassiz (1807–1873)—accepted Cuvier's views of the branching form of life but more explicitly interpreted this branching as a result of creation by a supernatural force over geological time (see chapter 4). Other followers, such as the English paleontologist Richard Owen (1804–1892), carried on Cuvier's tradition in comparative anatomy but as with Cuvier did not clearly reject or advocate an evolutionary basis for the origin of species.

Others such as Jean-Baptiste Lamarck, the French naturalist and evolutionist and a mentor of the younger Cuvier but later adversary, advocated a *scala naturae* or single-series view of the history of life earlier in his career. According to R. W. Burkhardt (1980), by 1802 Lamarck indicated that animal species could not be arranged linearly but rather formed "lateral bifurcations," and further that by 1815 Lamarck viewed a single line of increasing complexity as untenable. By at least 1809, Lamarck allowed, if not strongly supported, ramifying histories of life (Archibald 2009).

Even as the relations between Cuvier and Lamarck soured, both men, whether by mutual influence or not, abandoned Aristotle's view of the *scala naturae* in favor of a more tree-like branching of life—for Lamarck evolution was the cause, whereas for Cuvier the reasons were more nuanced. With the profound influence of such scientists, the *scala naturae* would seem doomed, and in large measure it was, but the *scala naturae* or great chain of being simply would not and will not die. It kept recurring in the nineteenth, early twentieth, and even twenty-first centuries.

Scala Naturae: Metaphorical Imagery That Will Not Die

Throughout the nineteenth century, the ramifying view of life and its representation took root under the aegis of both evolution and creationism within the expanding fields of biological sciences, but the great chain of being never really went away, notably in some philosophical circles (for example, Grindon 1863), in the public's eyes, or even in the work of some scientists. For example, the Scottish geologist Hugh Miller (1802–1856) produced what we would call fossil range charts in *Testimony of the Rocks* (1857), but with several twists (Archibald 2009). There are three such diagrams—one for plants, one for animals, and one specifically for fishes. Geologic time is shown, but rather than the traditional upward march of time, Miller's geologic time goes from oldest at the top to youngest at the bottom. His fossil range charts present us with simple, straight lines for each major group, yet surprisingly he calls each of the three diagrams a genealogy. For us, the only way in which such diagrams could be construed as a genealogy is if we restrict the term to each line representing the history of a particular group rather than any sense of connection of the groups. From Miller's description of these multiple simple lines, it becomes clear he has some sort of multiple *scala naturae* in mind: "The chain of animal being on its first appearance is, if I may so express myself, a threefold chain;—a fact nicely correspondent with the further fact, that we cannot in the present creation range *serially*, as either higher or lower in the scale, at least two of these divisions" (45–46). Miller's rather odd representations become clearer when his antievolutionist views are noted.

Unlike the obvious intent of scientists such as Miller, others inadvertently fanned the resurgence of the *scala naturae* and the idea of progression in nature. One of the most infamous examples appears as the frontispiece for Thomas Henry Huxley's (1825–1895) book *Evidence as to Man's Place in Nature* (1863), drawn after work by the natural history artist Benjamin Waterhouse Hawkins (1807–1894). In the pop-iconic illustration, we see from left to right a gibbon, an orangutan, a chimpanzee, a gorilla, and a human (figure 1.6*A*). Except for the slight forward tilt of the three central figures, all stand in an essentially upright posture natural only to the bipedal human. Huxley likely did not intend to suggest a progressive succession from gibbons through to man, but to place humans within the context of other primates.

The choice of presenting the other apes in a nearly bipedal human posture undoubtedly evoked this idea of relationship if not progression. Nevertheless, the result is the impression of progression from the "lower" apes through to "higher" humans. This message echoes in a passage in the chapter "On the Relationships of Man to the Lower Animals." Huxley (1863) writes that because differences in embryos of humans, apes, and dogs diverge only quite late in development and that

startling as the last assertion may appear to be, it is demonstrably true, and it alone appears to me sufficient to place beyond all doubt the structural unity of man with the rest of the animal world, and more particularly and closely with the apes. Thus, identical in the physical processes by which he originates—identical in the early stages of his formation—identical in the mode of his nutrition before and after birth, with *the animals which lie immediately below him in the scale*—Man, if his adult and perfect structure be compared with theirs, exhibits, as might expected, a marvellous likeness of organization. He resembles them as they resemble one another—he differs from them as they differ from one another. (83)

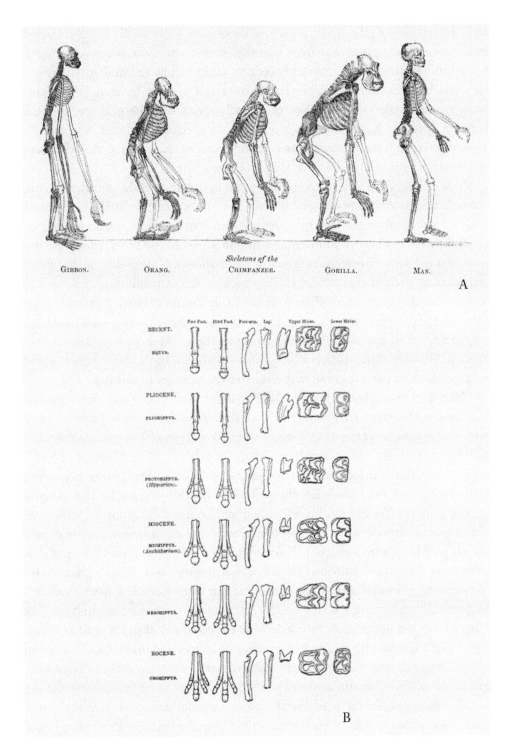

FIGURE 1.6 (*A*) Frontispiece of skeletons from Thomas Henry Huxley's *Evidence as to Man's Place in Nature* (1863); (*B*) Othniel Marsh's "Genealogy of the Horse," from "Polydactyl Horses, Recent and Extinct" (1879).

Huxley's words unquestionably emphasize the unity of humans, apes, and other animals, but whether intended or not, the phrase that I italicized in this passage indicates he views that there exists a scale in nature with humans at its apex.

The implications of the frontispiece illustration and passage from Huxley's text were not lost on his critics, some of whom were quite severe in their response. The eighth duke of Argyll (George Douglas Campbell, 1823–1900) stated, "On the frontispiece of this work he exhibits in series the skeletons of the Anthropoid Apes and of Man. It is a grim and a grotesque procession" (Campbell 1867:284; quoted in Desmond 1994). As is clear from his other statements in the same work, Argyll knew full well that Huxley explicitly only wished to show the similarity between apes and humans. But unlike Huxley, who wanted to place humans in the same order of mammals as other primates, Argyll argued that the gulf was so great between apes and humans that not only should humans be placed in a separate order of mammals but humans should be in their own class, akin to the other land vertebrate classes—Amphibia, Reptilia, and Aves.

This difference of view as to humans' place in nature paled in comparison with the scientific row Huxley engaged in with the English anatomist Richard Owen, who wished to place humans in a separate order from other primates based on small differences between humans and apes in a part of the brain, which Huxley soon showed was a misrepresentation of the anatomy. Nonetheless, it was characterizations such as that by Argyll that permeated the public psyche as well as some scientific circles as to where humans fit in nature. If Huxley presented any other sort of progression to the exclusion of humans, the response would have been negligible, but of course Huxley was trying to show where we rested in nature's scheme. While no great chain of being, this illustration had a clear impact on the perception of human evolution that erroneously continues to this day.

Next to humans, horses provide some of the richest history and lore surrounding the rise of evolutionary thought (MacFadden 1992). Huxley became involved with this as well, most famously because of a visit in 1876 with the American paleontologist Othniel Charles Marsh (1831–1899) at Yale's Peabody Museum of Natural History. Marsh is reported to have produced boxes of fossil horse bones and teeth dating from the middle Eocene to the Recent. This seeming linkage of equid species from the Eocene *Orohippus* (the earlier "Eohippus" was not yet known) to *Equus* greatly impressed Huxley as a marvelous proof of evolution. Huxley drew a sketch for Marsh with a fanciful "*Eohomo*" riding a still conjectural "*Eohippus*" that survives at Yale. Marsh provided Huxley with a quasi–*scala naturae* diagram showing subsequent evolutionary steps of the feet and teeth ascending from *Orohippus* to *Equus* (Desmond 1997), which Huxley (1877) soon used in a lecture in New York City. Marsh did not publish this scale of horse evolution diagram until 1879 (see figure 1.6*B*). Unlike Huxley's earlier frontispiece of apes and man, Marsh's diagram unmistakably shows a ladder-like evolution of horses—a great chain of horse evolution. As discussed by Bruce MacFadden (1992) in his book on horse evolution, not all paleontologists followed Marsh's lead for a straight ascent of horses through time. Soon after the turn of the century, more tree-like branching diagrams began to appear, some of which will be explored in a later chapter.

With *On the Origin of Species* (1859), Charles Darwin (1809–1882) made evolutionary ideas acceptable to the scientific community, if not quite so among the public, but his theory of natural selection rested on shakier grounds among scientists. By the turn of the twentieth century, evolution by means of natural selection was a theory in crisis for several reasons. One of note was the discovery of Gregor Mendel's work and the birth of genetics, which sought to displace natural selection as the cause of evolution as well as the source of genetic variation. Another reason relates to the resurgence of Lamarckian ideas of evolution along with other proposed evolutionary mechanisms under various names such as orthoselection, orthogenesis, and aristogenesis.

George Gaylord Simpson (1902–1984), perhaps the greatest evolutionary paleobiologist of the twentieth century, did not support these various orthogenetic hypotheses. In 1944, he wrote, "Most theories of this school, however, involve an element of predestination, of a goal, a perfecting principle, whether as a vitalistic urge, or a metaphysical necessity, or a frankly theological explanation of evolution according to which it is under divine or otherwise spiritual guidance" (152). At one end of the stronger theological spectrum for orthogenesis sits the French priest-paleontologist Pierre Teilhard de Chardin (1881–1955), whereas at the other end we find the American paleontologist Henry Fairfield Osborn (1857–1935). Osborn (1933, 1934) claimed that his version, aristogenesis, lacked any metaphysical, predeterministic, or perfecting principles, but as Simpson (1944) pointed out, this is a matter of definition because Osborn noted that aristogenesis "is definite in the direction of future adaptation" (152) and the word means "to bring into being the best of it kind" (152). Whatever the intention, Osborn's and others' similar ideas resulted in the portrayal of evolutionary modifications as occurring along a single line of change, hence the resemblance of the older *scala naturae* or great chain of being. Osborn's many illustrations along with his text strongly enforce this relationship between imagery and ideas.

Figure 1.7 presents a few iconographic images over Osborn's career, but it must be emphasized that he was far from being alone in his views; further, while the images discussed here epitomize his ideas on aristogenesis, he and those who worked for him produced many branching tree-like diagrams as well. This said, his orthogenetic imagery still reverberates. His 1929 monograph on titanotheres (relatives of horses, tapirs, and rhinos) is replete with progressive evolutionary images such as that in figure 1.7*A*, showing the frontal and side head views with ever larger and more complex nose horns illustrating, as he writes, "the progressive stages of development."

Certainly the most iconographic images of orthogenesis, second only to that of human evolution, or perhaps a fish emerging from the sea, are those showing the 55-million-year straight march in horse evolution from "*Eohippus*" to *Equus* discussed earlier (Osborn 1905; see figure 1.7*B*). Our perceptions do change; it is the nature of scientific theories. The American Museum of Natural History in New York, which Osborn once directed, has one of the largest collections of fossil horses in the world. Its exhibits document well what we thought we knew and what we think we now know about horse evolution. In the front part of the museum's display, horse evolution shows "a steady progression along a single pathway—until recently a widely held view of evolution. Here the horse is seen to evolve in a neat, predictable line, gradually getting larger, with fewer toes and longer teeth." But behind this, the exhibit also presents a more current scientific view of horse evolution showing it "to be a more complex, branching history."

Figure 1.7*C* comes from Osborn's (1936, 1942) monumental two volumes on proboscideans. It combines a more tree or starburst pattern but still for the most part shows the steady march in each lineage of elephantoids. In figure 1.7*D*, from a paper published by Osborn in 1933, the lower (grinding) molars of mastodons

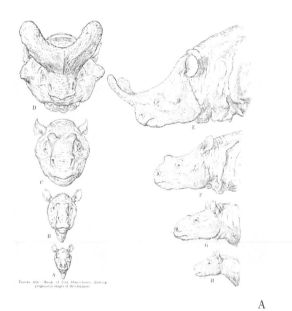

A

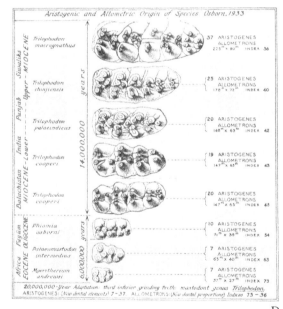

B

C

D

FIGURE 1.7 Henry Osborn's (A) "Progressive Stages of Development" of titanotheres, from *The Titanotheres of Ancient Wyoming, Dakota, and Nebraska* (1929); (B) "The Evolution of the Horse," from "Origin and History of the Horse" (1905); (C) "Adaptive Radiation from Genus *Elephas*," from *Proboscidea* (1936); and (D) "Aristogenic and Allometric Origin of Species" for mastodon molars, from "Aristogenesis, the Observed Order of Biomechanical Evolution" (1933). ([A] courtesy of the U.S. Geological Survey; [C] and [D] courtesy of the American Museum of Natural History)

ascend in an Aristotelian fashion, from small and simple at the bottom to large and complex at the top. The diagram even provides measures of Osborn's imagined aristogenes.

Steve Gould's Bane

The *scala naturae* continues to the present day in any number of guises. While not inhabiting any serious works intending to elucidate evolutionary relationships, it is alive and well in advertising and satirical humor, usually portraying a progressive march of fish coming out onto land, or a march of ape-like creatures to humans. Although these *scala naturae* are intended to hawk a product or skewer an adversary, underlying them all is our inability to jettison this as the iconography of evolution.

I am not certain when and where the current usage of the fish-to-mammal progressive march arose, but for the images of smaller apes to humans, I am quite certain that we can pinpoint the most recent source, if not necessarily the origin. The source of the ape-to-human transition originates in the popular book *Early Man* (Howell 1965). The illustration in question encompasses parts of five foldout pages, each page measuring 8 by 11½ inches (20 by 29 cm) (figure 1.8*A*). It is quite impressive, but I was taken aback because I knew both the author and the illustrator: the author, the American anthropologist F. Clark Howell (1925–2007), served on my doctoral examination committee at the University of California, Berkeley, and the artist, the Pulitzer Prize–winning, Russian-born American nature artist Rudolf Zallinger (1919–1995), I knew from my time on the faculty of Yale University. The illustration, "The Road to Homo Sapiens," shows fifteen species or varieties stretching back some 22 million years. The first sentence in the caption asks, "What were the stages of man's long march from apelike ancestors to *sapiens*?" (Howell 1965:41). I am quite certain that Howell would not have argued that human evolution was literally the straight line of ascent depicted in Zallinger's nearly fifty-year-old illustration, but I am equally certain that many, if not most, people did and do perceive human evolution to be this inevitable progressive march forward to humankind.

This kind of representation fuels profound misunderstandings by the general public of how evolution operates. It was a favorite topic of the American paleontologist Stephen Jay Gould (1941–2002). He criticized the view showing the history of life as a great chain of being progressing from simpler to ever more complex organisms. Rather, Gould argued that evolution is a process of diversification in which no progress can be detected. While the issue of progress in evolutionary change remains a more controversial topic, few biologists would disagree that the history of life has been one of diversification. Combining a somewhat lighthearted approach with a serious intent, Gould (1989, 1991, 1993) often wrote and lectured about our misperceptions of evolution as a straight-line, progressive process using Zallinger-inspired images in advertisements and political satire. The images in figure 1.8*B–E* are four of my favorite examples, with no further commentary required.

A

Somewhere, something went terribly wrong

B

C

D

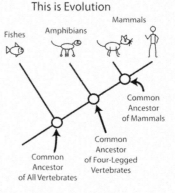

E

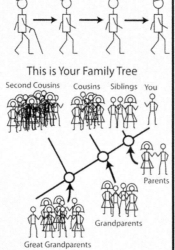

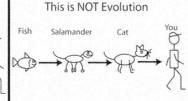

F

In figure 1.8*F*, an equally self-explanatory cartoon goes a long way to replacing Aristotle's ancient *scala naturae* with a much more current metaphor of how biologists perceive of modification with descent. This more accurate representation is often referred to as a tree of life, a metaphor that is even more culturally ubiquitous than the ladder of life. Yet the pedigree of the ladder metaphor is quite easily traced in Western culture, but not so that of the tree. The tree image can be easily identified in ancient Western sources whence our ideas of a biological evolution arose, but how the tree morphed from religious to biological symbolism is harder to trace.

FIGURE 1.8 (*A*) Rudolph Zallinger, "The Road to Homo Sapiens," from F. Clark Howell's *Early Man* (1965); (*B*) "Somewhere, Something Went Terribly Wrong"; (*C*) Bill Day, "Nobel Scientists Discover the Missing Link," *Detroit Free Press*; (*D*) untitled cartoon; (*E*) Larry Johnson, untitled drawing, *Boston Globe*; (*F*) Matthew Bonnan, "The Family Tree—Not Family Tree and Evolution—Not Evolution" (2010). ([*A*] reproduced with permission of the Zallinger family; [*B*] and [*D*] unattributed Web source; [*C*], [*E*], and [*F*] used with permission of the artists)

CHAPTER TWO

The Roots of the Tree of Life

A tree constitutes the single most powerful and most often used image of evolutionary history. Unlike the easily traceable Aristotelian ladder or scale of life, the origins of the biological tree of life imagery present a much more tangled but nonetheless traceable history. Evolutionary tree of life imagery constitutes an amalgam of male descent traceable from at least the Roman Republic with combined Roman religious and political acanthus-laden, tree-like imagery borrowed from the Greeks that appeared in temples recording birth-death-rebirth within cyclical nature. Early Christians adopted the acanthus/tree motif almost wholesale, with increasing dollops of Christian symbols added over time. Some truly phantasmagorical Christian tree designs and mystical images appeared. Within this framework emerged a hybrid of religious trees and familial trees using the ancient ideas of tracing familial descent within the context of a tree-like ascent.

Roman Obsession with Ancestors and Reverence for Nature

Ancient Romans at once feared and venerated their dead. After cremation, internment took place outside the city walls, except in the case of a few lucky emperors. Yet any Romans who could afford to do so prominently displayed wax masks (*imagines*) of their ancestors in the atria of their houses along with a stemma centered on the lineal descent of the male heirs of the household. The word "stemma" borrows from the Greek word for "wreath," and in the plural form "stemmata" means "wreaths or garlands of honor" but also refers to pedigrees and lineages. Italians still call their family coat of arms their stemma, but whereas some sculptural representations of ancestral masks exist (for example, a statue of a patrician holding lifelike *imagines* of his grandfather and father [Museo del Centrale Montemartini, first century B.C.E.]), no stemmata survive.

A number of ancient Roman sources describe both the masks and the stemmata. For example, in the first century C.E., Pliny the Elder (1952) writes, "In the halls of our ancestors . . . wax models of faces were set out each on a separate sideboard . . . to be carried in procession at a funeral in the clan . . . pedigrees too were traced in a spread of lines running near the several painted portraits" (book 35.6). Seneca (1935) noted around the same time, "Those who display ancestral busts in their halls, and place in the entrance of their houses the names of their family, arranged in a long row and entwined in the multiple ramifications of a genealogical tree—are these not notable rather than noble?" (book 3.28.2).

Stemmata probably resembled later forms in which wavy lines or ribbons connected names or portraits of forebears, often in medallions starting at the top with the common ancestor, sometimes expanded to show filial relationships, women in the family, or even adoptions (Klapisch-Zuber 1991, 2000). Maurizio Bettini (1991) emphasizes that these stemmata never showed tree-like genealogies, notably because they were read from top to bottom, not upward as in a tree, but on occasion references to *ramusculi* (branchlets) of the stemmata occur for various individuals, a feature that later became important for jurists in describing kinship (Klapisch-Zuber 1991).

Although the Roman dead were mostly placed outside the city walls, the city streets were lined, arched, and diverted by great heroic monuments as well as all manner of political and religious edifices. In ancient Rome, politics and religion formed an inseparable bond, even as they lamentably do today in many parts of the world. A small, now lesser known but in its day very powerful politico-religious structure arose during the reign of Augustus Caesar on the Campus Martius (Field of Mars), halfway between the still splendid Pantheon (rebuilt by Hadrian in 126 C.E.) and the now derelict mausoleum or tumulus of Augustus Caesar. This structure, the Ara Pacis Augustae (Altar of Augustan Peace), consecrated in 9 B.C.E., celebrated the emperor's triumphs.

The Ara Pacis provided the stuff for ancient Roman texts both in Rome and around the empire. The *Res gestae Divi Augusti* (*Acts of Divine Augustus*) engraved by the Galatians on the walls of the temple of Augustus in what is modern-day Ankara, Turkey, references the Ara Pacis: "[T]he senate resolved that an altar of the Augustan Peace should be consecrated next to the Campus Martius in honor of my return" (Velleius Paterculus 1924:12). The historian Dio Cassius (1917:54.25, 3) wrote about discussions as to whether the altar should be built in the Curia (Senate house); Ovid (1931:1.709) made reference to the Ara Pacis in several works, for example, *Fasti* (*The Festivals*); and the altar is depicted on the reverse side of ancient coins such as one with Nero and another with Domitian on the obverse side (for other sources, see Rossini 2008). Another, less certain image of the Ara Pacis forms part of the rectangular center of a possibly early-fourth-century hanging mosaic now housed in the Room of Masks in Rome's Palazzo Colonna dealing with the founding of the city. In the central rectangle, the goddess Roma appears as Minerva; also present are the she wolf with Romulus and Remus, and their discoverer, the shepherd Faustulus. In the upper right of the

A

B

C

mosaic appears an obviously rectangular structure with a pedestalled base and triangular cornices as well as a semicircle arising in the middle (for example, Safarik 2009). If any surface detail ever existed, age and wear obliterated it. This may be an early-fourth-century remembrance of the Ara Pacis, as palazzo guides indicate, or perhaps a sarcophagus.

Fragments of the lost altar appeared in the sixteenth century and formed the basis of one or more engravings (figure 2.1C), but these fragments then once again vanished (Vickers 1975). The true rediscovery and recognition of the Ara Pacis dates from only the last part of the nineteenth and the beginning of the twentieth century. With great difficulty, many fragments were extracted from underneath existing buildings in the 1930s and relocated, reassembled, and restored near the Mausoleum of Augustus. In 1938, Mussolini had a glass-sided enclosure built around the Ara Pacis as part of his Piazza Augusto Imperatore glorifying Fascist Italy. A new enclosure dating from 2006 now protects the Ara Pacis (Crow 2006).

The exterior of the Ara Pacis appears today as a rather austere off-white marble structure (see figure 2.1A), but when first built, its bas-relief walls shown brightly with color—certainly a striking presence on the open Campus Martius even though it measures only 35 feet (10.6 m) wide by 38 feet (11.6 m) long, and 20 feet (6.3 m) high (Rossini 2008). The longer sides, both open in the middle, had faced east and west, with the west being the entrance. The most famous section of the Ara Pacis encircles the exterior upper half of the altar, showing, among others, a procession of members of Augustus's family as well as more allegorical human and animal figures (see figure 2.1B). The lower half, sometimes referred to as the vegetal frieze, also encircles the structure. It is the largest such vegetal frieze known from ancient Greece or Rome. The exquisitely crafted, lifelike imagery suggests the work of Greek artisans (Hughes 2011).

Figure 2.2A shows the north side of the vegetal frieze, with an enlarged section of the north frieze immediately below (see figure 2.2B). An exhibition of the vegetal frieze in the Ara Pacis Museum noted about seventy species of plants in the frieze, although Caneva (2010) now identifies ninety unique plant species. This number will likely grow with more discoveries and analysis. The common plants include bear's breeches (*Acanthus*), Italian arum (*Arum*), lily (*Lilium*), date palm (*Phoenix*), water lily (*Nymphaea*), and above all asters and thistles (Carduaceae), most representing plants found in local meadows, grazing lands, and scrublands typical of the Mediterranean region.

Although sometimes treated simply as a decorative element of the Ara Pacis, the vegetal frieze clearly denotes symbolic themes. The plants form a continuous linkage with the dominant acanthus, a symbol of immortality and resurrection (Caneva 2010). According to the exhibition at the Ara Pacis Museum, the frieze evokes the idea of self-replicating clones and germination during the summer

---

FIGURE 2.1 Ara Pacis Augustae: (*A*) exterior; (*B*) procession of members of Augustus's family, on the exterior; and (*C*) photograph of an engraving of a fragment from the Ara Pacis attributed to Agostino Veneziano (ca. 1530–1535).

---

or following a drought. The profusion of thistles unfolding and blooming, possibly symbolizing the Augustan *aurea aetas* (golden age), suggests this. The profusion of spiraled branches, leaves, buds, and flowers but no fruiting represents the endless natural cycle of regeneration rather than an ending, known from ancient Greek as *anakyklosis*, an entirely pre-Christian perception resembling Lucretius's (2006) statement from *De rerum natura* (*On the Nature of Things*), "nullam rem e nihilo gigni divinitus umquam" (1.150). This is variously translated as "Nothing from nothing ever yet was born" (Rossini 2008) or "Nothing at all was ever born by divine agency" (E. Genovese, personal communication, 2012). The conspicuous swans at the top of the frieze honor Apollo, Venus, or Augustan patron gods (see figure 2.2*C*). Other small animals appear throughout: scorpions, snails, lizards (see figure 2.2*D*), butterflies, snakes threatening nestling birds (see figure 2.2*E*), and frogs (see figure 2.2*F*).

For us today, the imagery of a tree to show genealogy or evolutionary history seems natural, but in pre-Christian Rome, top–down stemmata showed human genealogy, and the idea of evolutionary history was unimagined; for these Romans, the branching acanthus motifs represented nature's cycle of birth-death-rebirth. With the toppling of the Roman religion by Christianity during the reign of Constantine in the early third century, Christianity began to embrace acanthus motifs in its churches, adding first simple Christian symbols and later magnificently ornate images. The vegetal friezes in Roman structures such as the Ara Pacis formed the ancestral basis for these later representations. Early Christians, however, began doing something different. Instead of showing top–down stemmata with lines of descent for the divine and other biblical personages, they arranged such figures from bottom to top, ascending to heaven superimposed on acanthus or tree-like backgrounds.

Expropriating the Acanthus Motif with Added Symbolism

Although now rather faded and forlorn, a mosaic from the fifth century bridges the time between the Ara Pacis of the first century B.C.E. and later, much better preserved and better-known Christian acanthus motifs. This mosaic occupies the apse in the narthex, or antechamber, of the octagonal Baptistery of the Arcibasilica Papale di San Giovanni in Laterano (Papal Archbasilica of St. John Lateran), likely the oldest baptistery in Christendom (figure 2.3*A* and *B*).

The Lateran Baptistery arose from an older Roman structure, possibly a nymphaeum donated by Emperor Constantine soon after his politically expedient conversion to Christianity and arguably before the Ara Pacis faded from view and memory. As with the Ara Pacis vegetal frieze, the narthex mosaic, although

FIGURE 2.2 Ara Pacis Augustae: (*A*) photograph of the north side of the color-enhanced (now black and white) vegetal frieze, (*B*) with an enlarged section. Some animals on the exterior of the Ara Pacis include (*C*) a conspicuous swan, (*D*) a lizard (broken head), (*E*) a snake threatening nestling birds, and (*F*) a frog (missing head). (The figures of the animals were enhanced for viewing.)

A

B

C

D

E

F

A

B

C

D

incomplete, shows some fairly crude but honest restoration (see figure 2.3*B*). The mosaic bears relatively few clearly Christian symbols. Flowers represent only three or so different forms (see figure 2.3*C*). Six small crosses, barely discernible except for one reconstructed on the far right (see figure 2.3*D*), encircle the top of the vegetal frieze. Four small doves and a lamb crown the vegetal portion of the mosaic. No other animals occupy the mosaic. A much later Baroque cross and two cherubs jut outward from the lower center of the mosaic, partially obscuring the restored acanthus at the bottom center. The central stalk arises from the acanthus, with two intertwined vines forming four ovals, quite similar in design to such ovals in later Christian mosaics. Plants occupy at least the lower two ovals, reminiscent of the many plant species in the central stalk of the Ara Pacis vegetal frieze.

The acanthus motif reaches its most glorious excess in the small Basilica di San Clemente (Basilica of St. Clement), dedicated to the traditionally recognized fourth pope, Clement I, and one of the oldest churches in the city (figure 2.4*A*). The extant basilica, in what passes for one of the quieter corners of Rome but only some 985 feet (300 m) from the Colosseum, dates from probably the early twelfth century, although times for building on the site remain confused and confusing (Boyle 1989; Collegio S. Clemente and Gerardi 1992). Churches abound in Rome, but this one holds more than a few surprises. The site preserves as many as four levels, at least three of which encompass religious histories. The lowest layer includes the remnants of structures likely destroyed in the fire of 64 C.E. during the reign of Emperor Nero. Above this, parts of structures from the first century preserve areas for religious practices of early Christians as well as areas for the Mithraic religion, including the mostly enclosed sanctuary and surrounding rooms considered as areas for indoctrinating followers into that religion, named after the Persian god Mithras. As the Christian sect came to dominate, it took over this space when the Mithraic religion became illegal in 395. The third level is the first and larger Basilica di San Clemente, probably completed in the late fourth century and heavily damaged by fire during the Norman invasion in the eleventh century. The slightly smaller, existing basilica arose upon the fourth-century structure relatively soon after 1100 (Boyle 1989).

These older structures below San Clemente offer a fascinating glimpse of religion throughout Roman history, but they do not provide the most striking extant features of San Clemente. This honor belongs to the incredible twelfth- or thirteenth-century golden mosaic that covers the semicircular recessed apse behind the altar (see figure 2.4*B*). The inclusion of clearly third- and fourth-century motifs suggests that this may be a copy of a similar and probably larger mosaic in the earlier version of the basilica. The large crucifix in the middle of swirling curlicue plant tendrils, however, does not come from third- and fourth-century sensibilities but represents a more twelfth-century religious perception (Boyle 1989). The crucifix emerges from a clump of foliage at the bottom middle

FIGURE 2.3 Baptistery of the Arcibasilica Papale di San Giovanni in Laterano: (*A*) exterior; (*B*) mosaic in the apse in the narthex; and details of (*C*) flowers and (*D*) a cross.

FIGURE 2.4 Basilica di San Clemente: (A) exterior and (B) mosaic in the apse.

of the mosaic. The clump is an acanthus plant, which pre-Christian Greeks associated with cyclical nature and life enduring beyond the grave. This sort of motif in architectural design, known as spiraling *rinceaux*, refers to the later French term for this botanical form (Semes 2004), with an unmistakable resemblance to the same motifs in the Ara Pacis and Lateran Baptistery. Interestingly, the French *rinceaux* derives originally from the *ramusculus* discussed earlier, which is Latin for "small branch."

As in the Ara Pacis vegetal frieze, the San Clemente apse mosaic sends out curved acanthus tendrils of various plants from a large central acanthus. In the mosaic, the large crucifix dominates the center, whereas the Ara Pacis acanthus sprouts a large vegetative stalk from its center comprising six different plant species (Caneva 2010) (see figure 2.2*A* and *B*). Even the number, directions, and manner of joining of the spiraling vines in the mosaic and the frieze bear an uncannily striking resemblance. The apse mosaic also presents many animals as well as people: deer, peacocks, chickens, cattle, sheep, goats, dolphins, and doves, some with Christian symbolism. The alternating positions of the doves on the upright of the crucifix (see figure 2.4*B*) recall the vegetative stalk emerging from the acanthus in the Ara Pacis frieze.

The mosaic supposedly expresses a Byzantine influence (Gerardi 1988) in the Cosmatesque or *opus Alexandrinum* style. Such mosaics grace much of medieval Italy, especially in the vicinity of Rome. Cosmatesque takes its name from the Cosmati family, the thirteenth-century craftspeople who arranged ancient marble fragments into geometric inlays, including interconnected spirals. Possibly the San Clemente apse mosaic includes some Byzantine designs, but the great similarity in design and symbolism between this mosaic, the Lateran Baptistery mosaic, and the Ara Pacis vegetal frieze cannot be dismissed, although they span more than a millennium. The Christian symbols represent life and constant rebirth, but whereas the vegetal frieze shows Roman themes, some with a Hellenistic source, the apse and Baptistery mosaics place these within the context of a Christian credo. Further, all three structures are in close geographic proximity. Before Mussolini moved the Ara Pacis to its current location in the 1930s near Augustus's tumulus, it stood little more than 1 mile (1.6 km) from San Clemente and another half-mile (0.8 km) farther on to the Lateran Baptistery.

The Ara Pacis vegetal frieze, the Lateran Baptistery narthex mosaic, and the San Clemente apse mosaic quite literally as well as metaphorically depict trees of life. The later emerging scientific use of the tree of life symbolism most directly comes from biblical references and familial genealogies, but even these derive from older traditions in Roman and Greek antiquity. We cannot say that the first-century B.C.E. Ara Pacis frieze led directly to the fifth-century C.E. Lateran Baptistery mosaic, which in turn led to the likely twelfth-century San Clemente mosaic (figure 2.5), but they resemble one another too much for mere coincidence, observations in part made separately by others (Toubert 1970; Tcherikover 1997; Caneva 2010). As we shall see, other early examples of the spiraling acanthus motif strengthen and support this conjecture. Although a Roman quip, "Se non è vero, è ben trovato" (If it's not true, it's still a good story), tempers the desire to accept such coincidences, I think the evidence supports the truth of the matter as to the continuity through time of these motifs.

A

B

C

Everywhere, Swirling Acanthus

The likely relationship between the Roman vegetal frieze and the two Christian apse mosaics stands more clearly because of the ubiquity of the acanthus motif. Architecturally, acanthus plants are everywhere, from the ostentatious capitals of Corinthian columns, such as those adorning the Pantheon in Rome (figure 2.6), to the spiraling *rinceaux* gracing the interiors and exteriors of many structures from ancient to modern. If one looks even passingly at buildings with any classical Greek or Roman influences, the acanthus abounds. Examples, much less elaborate than in San Clemente and less integral to the design, survive in or near Rome and vary considerably in age.

The oldest example, and the only one not found in Rome, adorns the very well-preserved Maison Carrée in Nîmes, France. As with many ancient Roman buildings—most famously, the Pantheon in Rome—it survives in such good condition because of its requisition by the Catholic Church. Built around 16 B.C.E. for an uncertain purpose, the building was dedicated between 5 and 2 B.C.E. to Gaius and Lucius Caesar, the sons of the building's designer, Marcus Vipsanius Agrippa, and heirs of Augustus Caesar. Above the columns supporting the building, a relief band of spiraling stems and leaves emerges from a central acanthus plant. The stems and leaves are interspersed with flowers and occasional birds (figure 2.7*A*).

The Terme dei Sette Sapienti (Baths of the Seven Sages) from 130 C.E. at Ostia Antica, the ancient port some 12 miles (20 km) southwest of Rome, preserves much of a circular mosaic floor 40 feet (12 m) in diameter. The circular room possibly served for some time as a market and at other times as the cold room of a bathing complex. From what can be determined from observation and various sources, two acanthus stalks arise from two quadrants 90 degrees from each other, each giving rise to spiraling acanthus (see figure 2.7*B*). Although some flowers are shown, a hunting scene with humans and a host of wild beasts among spiraling acanthus dominates the design. Notably, as we pass from the first-century B.C.E. Ara Pacis to the second-century C.E. Terme dei Sette Sapienti to the twelfth-century San Clemente, animals and humans become more prominent in the design as the likely meaning of the design changes.

The next younger example preserves a painted decoration from a room known as the Sala dell'Orante found in a fourth-century Roman house supposedly belonging to two men, John and Paul, who worked for Emperor Constantine. As the story goes, they suffered martyrdom, and the Basilica di Santi Giovanni e Paolo al Celio (Basilica of Sts. John and Paul on Caelian Hill) now overlies their house. A painted band in the Sala dell'Orante, although faded and less elaborate than the relief on the Maison Carrée and the mosaic at Ostia Antica, unquestionably shows the same motif of an acanthus plant, spiraling stems, and leaves (see figure 2.7*C*).

FIGURE 2.5 Comparison of (*A*) mosaic in the apse of the Basilica di San Clemente, (*B*) mosaic in the apse of the Baptistery of the Arcibasilica Papale di San Giovanni, and (*C*) enlarged section of the north side of the vegetal frieze on the exterior of the Ara Pacis.

FIGURE 2.6 Top of a Corinthian column on the Pantheon.

A number of churches in Rome use the acanthus as a decorative motif. The oldest of these examples is almost a thousand years younger than the painted acanthus decoration in the Sala dell'Orante. These churches vary in the elaboration of both painted decoration and ornamented relief. The order of the often overlapping time intervals when most of the extant portions of these basilicas were built or repaired is Sant'Agostino (St. Augustine) in the thirteenth through fifteenth centuries (figure 2.8*A*), San Luigi dei Francesi (St. Louis of the French) in the sixteenth century (see figure 2.8*B*), and Santa Maria del Popolo (Our Lady of the People) in the fifteenth and sixteenth centuries (see figure 2.8*C*). Just as with the older Maison Carrée ornamentation and the Sala dell'Orante decoration, all these later examples of the acanthus motif do not represent the centerpiece of any display, altar, nook, or nave but serve as embellishments that frame paintings or form wall or ceiling panels.

Santa Maria Maggiore (St. Mary Major) uses acanthus plants as a more important but still not central theme in its interior decoration. This major basilica in Rome supposedly ranges in age from about 350 c.e. to the eighteenth century. The acanthus plants in question flank each side of the main apse mosaic, which depicts the Coronation of the Virgin, created by Jacopo Torriti in the late

FIGURE 2.7 (*A*) Spiraling stems and leaves emerge from a central acanthus plant at the top of the exterior of the Maison Carrée, with details of birds; (*B*) partial image of the mosaic floor at the Terme dei Sette Sapienti; (*C*) illustration of painted decoration of acanthus from a room known as the Sala dell'Orante, from Germano di San Stanislao's *La Casa Celimontana dei SS. Martiri Giovanni e Paolo* (1894).

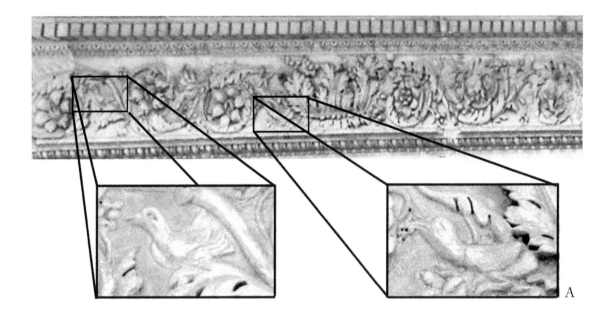

A

B

C

A

B

C

thirteenth century. The tree-like acanthus plants flank the far left and right sides of the main panel of the mosaic, each sending forth six large, spiraling stems (figure 2.9*A*). The center of each spiral bears a flower, each of which appears to be distinct from the others. At least twelve smaller spirals on each side possess different flowers. In addition, eleven birds adorn the (viewer's) left acanthus and nine embellish the right, including—among others—ducks, peacocks, cranes, and a raptor, as well as small passerines (see figure 2.9*B*).

As an aside, having worked on mammals throughout my academic career, I take special interest in the only two mammals represented in the mosaic. A rabbit sits rather boldly on a flower-like blossom on the lower-right spiral of the right acanthus. A small rodent scurries up the lowest small spiral on the left (see figure 2.9*C*). I cannot determine what species this might be or what, if anything, it may symbolize. A likely candidate, the black rat (*Rattus rattus*), reached the Roman Empire from Asia at least by the first century C.E. The long, thin tail suggests this species, but it also might represent the edible or fat dormouse (*Glis glis*), much favored by Romans as a snack. This species and that portrayed in the mosaic are rather chubby, but the dormouse has a fluffier tail. Notably, one of the frescos in the fourth-century home below the Basilica di Santi Giovanni e Paolo al Celio also shows a rodent more strikingly similar to the edible dormouse.

These acanthus and spiraling *rinceaux* motifs, arguably with the exception of those in Santa Maria Maggiore, largely represent subsidiary embellishments. Not so for the San Clemente apse mosaic, which far exceeds the ornamental and decorative acanthus motifs (see figure 2.4*B*). The San Clemente acanthus spirals intertwine with many layers of religious symbols and symbolism; one description intones that "the sum total is that the mosaic depicts the Tree of Life, which has withered after sin entered the world but then turns into the new Tree of Life, the Cross of Christ which is the ultimate symbol of Christ's victory over sin" (Churches of Rome 2012). The use of acanthus in the basilica obviously continues the Greek ideas of life beyond death into the Christian notions of resurrection but on a far more elaborate scale.

The Tree and the Acanthus Are Joined

The swirling acanthus trees in San Giovanni in Laterano and San Clemente, with their Christian symbolism of the crucifix, evoke the life of Christ yet provide no literal genealogy. The crucifix dominates the form in San Clemente, whereas that in Santa Maria Maggiore celebrates the Coronation of the Virgin (see figure 2.9*A*). This is not so with other acanthus trees of similar vintage. In these other trees, a true blending of acanthus and genealogy occurs, albeit of biblical persons, and rather than the descending in a human lineage, they ascend to heaven—an obvious reversal of the stemma.

FIGURE 2.8 Scrolled acanthus motifs in Roman churches: (*A*) Sant'Agostino; (*B*) San Luigi dei Francesi; and (*C*) Santa Maria del Popolo.

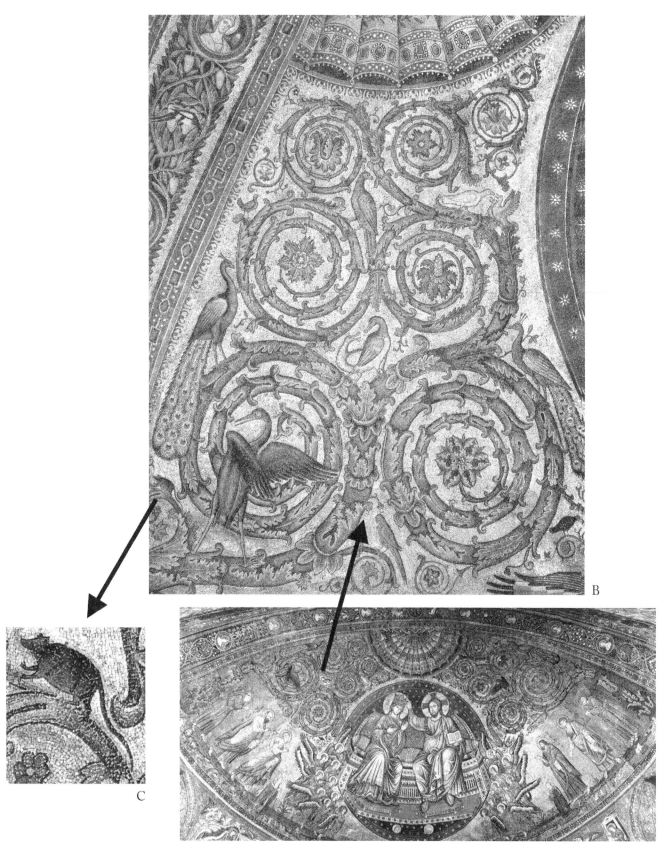

B

C

A

A common theme, the Tree of Jesse, such as in the beautiful window at Chartres Cathedral, most often shows Jesse as the base or root from which the tree leads through the House of David most often upward to Jesus of Nazareth. Inspiration for this image comes from the prophecy in Isaiah 11:1. The translation from Latin of the Vulgate Bible (commonly in use in the Middle Ages) reads, "A rod will come forth from the root of Jesse, and a flower shall rise up." The Latin word for "rod," *virga*, sometimes served as a pun referring to the Virgin Mary (*virgo Maria* in Latin), with the flower referring to Jesus. In the representations of the Tree of Jesse in churches and illuminated manuscripts, Jesse forms the root or base from which sprouts the tree ascending to the Virgin and Child. Various versions show assorted prophets and saints (figure 2.10).

In these trees, we see subdued and less subdued echoes of the acanthus motif in the Ara Pacis vegetal frieze and the Lateran Baptistery and San Clemente mosaics. One fifteenth-century carved example of the spiraling *rinceaux* Tree of Jesse comes from Germany. It appears on the south side of St. Lamberti in Münster. A tall portal includes a Tree of Jesse first carved in sandstone in the fifteenth century and then restored in the early twentieth century (see figure 2.10*A*). A splendid late Gothic carved example of the Tree of Jesse with the spiraling *rinceaux* ornamentation exists in the Cathedral of St. Peter in Worms, Germany. The north aisle preserves five tympana, the semicircular decorative surfaces over entryways, one a Tree of Jesse (see figure 2.10*B*). These tympana once surmounted the entrances of the demolished cloister associated with the cathedral from the end of the fifteenth century. Examples of the Tree of Jesse coeval with the San Clemente mosaic derive from an unknown source from Würzburg, Germany, from around 1240 to 1250 (Kren 2009) (see figure 2.10*C*), and the Capuchin's Bible, from around 1180 (see figure 2.10*D*). Both include the flourished curlicue vines entwining the edges of each scene, but despite their unmistakably tight coils, neither of these comes close to the exuberance of the Ara Pacis frieze or the San Clemente mosaic. Such vine motifs also occur as marginal embellishments on many other illuminated manuscripts, although none with a clear tree-like meaning as in Trees of Jesse.

Examples abound, but the four shown in figure 2.10 most clearly evoke earlier Hellenistic/Roman and then Christian trees of life replete with symbolism harking back to the acanthus and its spiraling tendrils. Two elements missing or nearly so are flowers and animals, except for one dog in the St. Peter tympanum and what appears to be a dove in the Capuchin's Bible, but they do ascend upward.

The Tree Mosaics of Otranto, the Sacred and the Profane

No discussion of trees of life in Christian churches would be complete without mentioning the unquestionably most grandiose of them—the 700-square-foot (65 sq. m) mosaic that covers the entire floor of the Cathedral of Otranto

FIGURE 2.9 Santa Maria Maggiore: *(A)* mosaic in the apse; *(B)* a left-spiraling acanthus; and *(C)* detail of an unidentified rodent.

A

B

C

D

(Nigro 2000; Gianfreda 2008). Otranto, a small town on the Adriatic Sea, lies almost at the end of the boot heel of Italy. The cathedral was consecrated in 1088, but the mosaic floor was not laid until the latter third of the 1100s, about a hundred years before the very different apse mosaic in San Clemente. Five distinct parts define the floor mosaic: the apsidal area behind the altar, the presbytery area in front of the altar, two small trees of life (one each in the side naves), and the large tree in the central nave (figure 2.11).

A tree motif does not define the apse floor mosaic behind the altar but shows a variety of biblical and phantasmagorical creatures. The presbytery floor in front of the altar does include a small tree at its base and sixteen circular medallions, some with clear biblical significance. The two medallions at the bottom flanking the small tree, presumably the Tree of Knowledge, show the temptation of Adam and Eve by the serpent. The medallion next to Eve shows the behemoth referred to in Job 40:15 ("Behold now behemoth, which I made with thee; he eateth grass as an ox" [King James Bible]), probably referring to a hippopotamus. The medallion just above Adam is a large, serpent-like form with a hare in his mouth; attributed to the leviathan in Job 41:1 ("Canst thou draw out leviathan with an hook?" [King James Bible]), it may refer to a crocodile (see figure 2.11B). In the upper left of the presbytery floor, medallions show the Queen of Sheba and King Solomon.

Other motifs in the presbytery may be satirical. Directly above the small tree but outside a medallion, a dog beats cymbals, symbolizing folly, and above it an ass strums a harp, symbolizing ignorance. Various phantasmagorical beasts are depicted. A siren with a split tail appears in one medallion. A centaur in another medallion fires arrows at the stag in the next medallion, with one imbedded in its chest. A griffin—sometimes mistakenly identified as a basilisk, cocktrice, or leopard—seems to be attacking a goat (another griffin is shown in figure 2.11C). A man, tentatively identified as the creator of the mosaic, kneels next to a unicorn. Real mammals include a camel, an elephant, a stag (with arrow), an antelope, a bull, and a leopard attacking a fox (representing lust and cunning, respectively).

The right nave holds the Tree of Redemption. It displays an identified biblical figure, Samuel, but also mythological creatures such as a harpy, a sphinx, a minotaur, and a lion, supposedly of Judah, biting a dragon. Names in Latin identify Abraham, Isaac, and Jacob in the left nave mosaic, which shows the final Resurrection and Judgment. Three beasts and a scapegoat guard the gates of hell, with a man bound hand and foot and a figure next to him identified as Satan.

When observed in plan view, the floor mosaic in the central nave bursts with detail. One does not know where to look first, but a few motifs stand out. A tree runs the entire length of the center of the nave, extending branches large and small. Twelve circular medallions near the top of the tree name the twelve months and show the zodiacal signs with human and animal figures performing normal

FIGURE 2.10 Tree of Jesse: (A) St. Lamberti; (B) Cathedral of St. Peter; (C) unknown source from Würzburg; and (D) Capuchin's Bible. ([C] reproduced by permission of the J. Paul Getty Museum, Los Angeles, Ms. Ludwig VIII 2, fol. 7v)

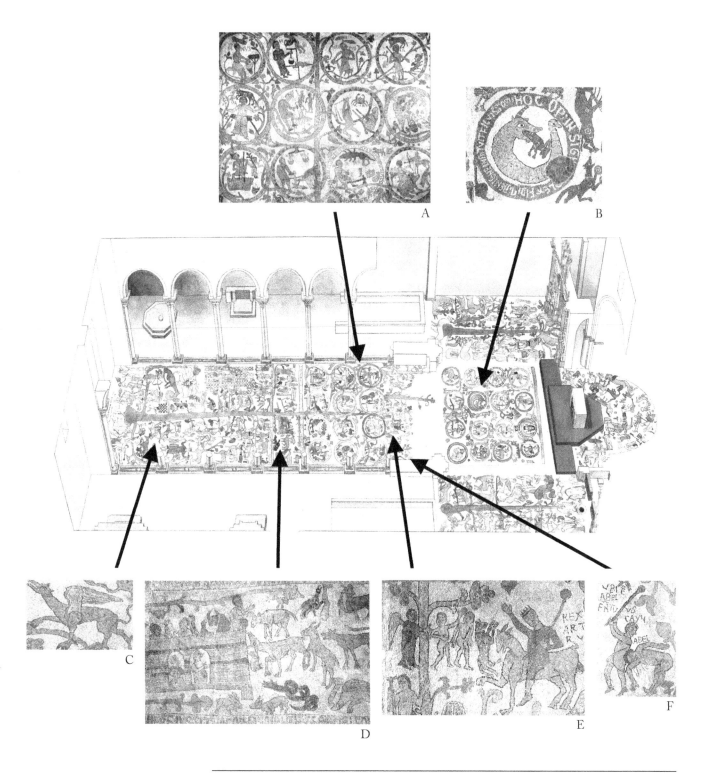

FIGURE 2.11 Mosaics of the Cathedral of Otranto, using a floor plan modified after Grazio Gianfreda's *Il mosaico di Otranto* (2008): (*A*) twelve circular medallions near the top of the tree name the twelve months and show the zodiacal signs with human and animal figures performing normal seasonal tasks; (*B*) a large, serpent-like form with a hare in his mouth, attributed to the leviathan in Job 41:1; (*C*) a griffin, sometimes mistakenly identified as a basilisk, cocktrice, or leopard, appears to be attacking a goat; (*D*) left to right, the story of Noah and the Flood unfolds; (*E*) Adam and Eve (*left*) partake of the Tree of Knowledge and (*right*) whimsically wave good-bye while being driven from the Garden of Eden by an angel, but a labeled King Arthur intervenes; and (*F*) labeled story of Cain and Abel's sacrificing to God, here showing Cain slaying Abel.

seasonal tasks (see figure 2.11*A*). Immediately above these and to the left, two small trees tell the story of Adam and Eve, first with their fall after partaking of the Tree of Knowledge. Next, to the right, they whimsically wave good-bye while being driven from the Garden of Eden by an angel. Oddly, a labeled King Arthur intervenes between this (see figure 2.11*E*) and the labeled story of Cain and Abel's sacrificing to God, here showing Cain slaying Abel (see figure 2.11*F*).

Immediately below the calendar medallions, reading from left to right, the story of Noah and the Flood unfolds (see figure 2.11*D*). On the next branches below, the eye is drawn to a checkerboard design on the left side that tells the story of the Tower of Babel. To the right, below, and diagonally across from the tower, a host of real and mythological humans, part-humans, part-animals, and animals occupy the bottom half of the mosaic. Immediately below the tower, the huntress Diana slays a stag. Slightly above and to her right, a clearly labeled figure of Alexander the Great stands before two griffins. At the very base of the expanded caudex of the tree stand two formidable elephants facing away from each other. In the mosaic immediately below and near the entrance to the church appears the name Pantaleone, the monk who supervised the work over the four years it took to complete the mosaic.

Pantaleone's intended meaning for this eclectic floor mosaic remains obscure. It certainly presents elements found in other Christian mosaics—primarily, Hebrew Bible and New Testament stories—but Greek mythology and even elements from Asia and the Islamic world often appear. Part of this may be explained by Otranto's proximity to these other regions. Various countries and peoples repeatedly invaded the Otranto region, and thus a cosmopolitan interpretation of the tree of life is not unexpected. The frequent inclusion of not only biblical but also more obscure texts within the mosaic seems out of place for a time when almost no one could read. In part it reads like an allegorical inside joke. Other trees in churches so far discussed do not register a clear sense of linear time. The Otranto mosaic as a whole provides an even more nonchronological narrative, although certain parts do follow a story line. Nevertheless, the tree design of the mosaic seems to be an attempt to unify many stories of the Judeo-Christian God with gods, humans, and animals, both imagined and real, from the cultures of the then known world. It is a history of sorts even if time frames and events thoroughly mingle with one another.

Cyclical Time Becomes Linear

From at least ancient Rome onward into medieval Europe, trees or other plant motifs became powerful religious symbols, many but not all opening upward toward heaven. Tree motifs in both Roman religions and Christianity used plants, especially acanthus, along with humans and a variety of real and imagined animals. In both traditions, the trees could symbolize death and renewal in nature or immortality of the soul. Cyclical themes of death and rejuvenation commonly occurred in these representations, with historically accurate representations of time of less concern, as especially seen at Otranto. Historical time versus cycles became more important parts of tree representations when a truer sense of spiritual and physical history emerged. At least two uses of history appeared at this time: one vision for the ages of man, as popularized by Joachim of Fiore, and the second employing trees to show human or even domestic animal genealogy. The ages of man framed a spiritual history for humans, whereas the human genealogical diagrams showed who begat whom.

Joachim of Fiore (1135–1202), an Italian monk, mystic, and biblical scholar, produced many sorts of geometric figures—circles in circles, triangles, spirals—but is best known for tree-like figures appearing in two primary works: *Liber concordie novi ac veteris Testamenti* (*Book of Harmony of the New and Old Testaments*) and *Liber figurarum* (*Book of Figures*) (Reeves and Hirsch-Reich 1972; Reeves 1999). Through both biblical study and mystical experiences, he often elucidated in these trees what he called the three stages of time. God's law defines the first stage. The second stage of grace comes through God's son. The third stage, coming in the future, forms from the Spirit with the time of liberty and love awaiting the Second Coming (figure 2.12). Joachim of Fiore envisioned these three stages as the tree of life mentioned three times in the book of Revelation. Representations of these trees often show intertwining branches, each of which encircles one of the three stages, emphasizing the three stages of spiritual history (see figure 2.12*A–C*). German romantics and their materialist and Marxist descendants inherited this, as did the positivist philosophy of Auguste Comte (Hestmark 2000).

Joachim of Fiore's revolutionary ideas conceived of the overlap of historical periods and change as a process of germination and fruition (Cook 1988). Notwithstanding claims that his trees express God's purpose through the natural time span of history in biological terms (Reeves 1999), he visualized a historical framework not common for his contemporaries; thus what better than to use a tree? A new shoot generates from those that preceded it and, in turn, germinates the ones that follow in an upward trajectory.

Most of his trees possess multiple layers of meaning, not always in plain view. One motif, his Tree Eagle, when viewed upright represents a tree-like form, but when inverted, an eagle's head reveals itself with branches for wings and tail (see figure 2.12*D*), supposedly a hidden symbol of the coming of his third historical age of the Spirit. Other, even more tree-like figures include trunks, branches, leaves, and buds but with the ever-present three-part spiritual history incorporated (see figure 2.12*E*). These trees presage and even remarkably resemble some of the biological trees of life that first appeared in the beginning of the nineteenth century, whether the budding profession of biologists attributed the cause of the change to a deity or to evolution.

FIGURE 2.12 Joachim of Fiore's (*A–C*) trees with intertwining branches that emphasize the three stages of spiritual history; (*D*) Tree Eagle, which viewed upright represents a tree-like form, but when inverted reveals an eagle; and (*E*) more tree-like figure but with the ever-present three-part spiritual history incorporated.

A

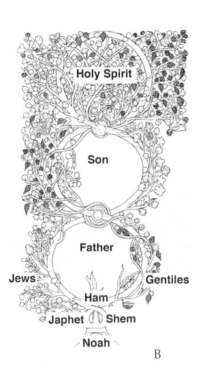

Holy Spirit

Son

Father

Jews

Gentiles

Ham

Japhet   Shem

Noah

B

C

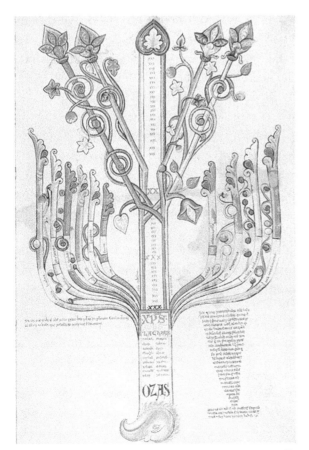

D

E

Placing the Tree Upright

From the tenth to the fourteenth centuries, tree-like genealogies proliferated, both profane and religious, such as the Tree of Jesse described earlier. A majority of these genealogical trees carried on the tradition of showing the idea of descent— that is, from the top down. By the latter half of the sixteenth century, the more botanical tree form came into vogue (Klapisch-Zuber 1991), but, as shown earlier, this tradition had worked its way into use through other, often religious sources showing trees opening upward toward heaven. A striking aspect of these representations occurred with the reversal from a top–down to a bottom–up genealogy with the adoption of a true tree image. Schemata transformed from the idea of descent, as Christiane Klapisch-Zuber (1991) phrased it, to the idea of "an ascent and spreading-out" (112). All the old meaning from which they arise meant that genealogies read from the top down, metaphorically as a stream flows. We now generally interchange freely the words "descent" and "ascent" in genealogic or evolutionary sense, showing the diagrams opening in various directions, but this came about relatively recently, perhaps in the past two hundred years at most.

The iconography on more than one occasion bordered on the absurd. Figure 2.13*A* and *B* shows trees that open downward, with ancestral roots topsyturvy. Figure 2.13*B* shows an upside-down Tree of Jesse. Even more confusion reigns when the tree finally becomes upright but the descent of one individual is shown at the base of the tree with ancestors in its upper branches. Figure 2.13*C*, a late-sixteenth-century family tree, at first appears to show a normal, tree-like genealogy, but look closer. The base of the tree shows Louis (Ludwig) III (1554–1593), known as Louis the Pious because of his religiosity, who served as the fifth duke of Württemberg from 1568 to 1593. Leafy branches, dangling fruit, overflowing cornucopias, and even spiraling architectural borders highly embellish this tree. The problem: the duke had no heirs, so why does he lie at the base of a family tree? The answer: the iconography is confusingly inverted. A genealogy of his ancestry or his pedigree is represented, starting with the sixth generation at the limb tips and extending down the tree, through his father, Cristoph (*left*), and mother, Anna Maria (*right*), to him at the base. Thus considerable confusion arises from using the same tree for both a pedigree and a genealogy. A pedigree shows all known ancestors of one person or a family and in older representations opens downward, whereas a genealogy shows a person's parents, the individual offspring, and all that person's descendants.

While visiting southern Italy in 2009, my wife and I stayed about 80 miles (130 km) north of Otranto in a *masseria* abandoned around World War II. *Masserie* are fortified farming estates in the southern coastal regions of Italy. Since Italy did

FIGURE 2.13 Trees that open downward: (*A*) ancestral roots facing up; (*B*) upside-down Tree of Jesse, from Vincent Placcius's *Justiniani Instiutiones juris reconcinnatae* (1682); and (*C*) what appears to be a family tree, but is actually a pedigree of Louis III, fifth duke of Württemberg. ([A] reproduced by permission of Innsbruck, Universitätsbibliothek)

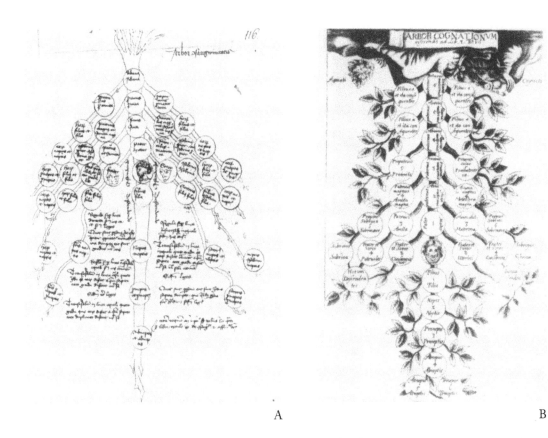

A

B

C

A B

FIGURE 2.14 (A) Badly water-damaged Italian family tree; (B) "Genealogical Tree of the Queen [Victoria] and Her Descendants."

not have a centralized government and armies until the nineteenth century, most *masserie* had to be defended against the various invaders that came ashore through the centuries. The one in which we stayed, Masseria Il Frantoio, preserves olive trees planted at least five hundred years ago. Although many of the buildings are from the nineteenth century, an underground olive press (no longer in use) may date from the sixteenth century. In the main house hangs a badly water-damaged family tree, which by appearances may date to at least the nineteenth century, if not earlier (figure 2.14*A*). I include it in spite of its poor condition because of its unusual form and presentation. A cartoonish, dark green-gray central stalk arises in the middle with now indiscernible dates and with side branches coming from the stalk. From the tree hang medallions with names and presumably birth dates, because the word *nato* (born) appears near the dates. *Nato* is also written on the stalk, but damage renders the dates illegible. The oldest discernible date on the lowest better-preserved medallion on the right reads 1639 or 1689. One can imagine, although not see, even older natal dates on the stalk. The most recent come from the early nineteenth century. Unlike most family trees, which show pedigrees of direct descent, the main stalk of this tree forms segments, with the medallions on side branches probably showing whole families.

A much more sophisticated (probably early-twentieth-century) version of a tree with medallions records the "Genealogical Tree of the Queen [Victoria] and Her Descendants" (see figure 2.14*B*). Unlike the Italian version, this one portrays Prince Albert, Her Majesty, their offspring, and in turn their offspring. This otherwise rather normal family tree represents the earlier generation at the base and later generations closer to the tips, much as we see with late-nineteenth- and twentieth-century representations of evolutionary history.

A particularly engaging tree-like representation comes from Lorenz Faust's quirky work *Anatomia statuae Danielis* (*Anatomy of Daniel's Statue*, 1585). He shows genealogical relationships not only as trees but also on the armored physique of a knight and another on the palm of a hand (Rosenberg and Grafton 2010). The tree shown in figure 2.15*A* traces the history of Saxon rulers. The tree evokes a rather massive presence as of a spreading oak, with each of the sparse larger limbs sprouting much smaller branches that march along horizontally in a Germanic fashion, budding off small heart-shaped leaves at regular intervals. Each leaf presents the name of a Saxon ancestor. As discussed in chapter 3, the tree first used to show botanical relationships in the earliest nineteenth century, although not evolutionary, employed the same motif—a tree with leaves but with the name of plant species rather than that of a Saxon ruler in each leaf.

Certainly one of the greatest uses of trees for biblical, royal, and pontifical lineages was that of Hartmann Schedel in *Nuremberg Chronicle* (1493). This monumental work provides many drawings of cities and real and imagined people, no matter from what epoch, dressed in medieval garb. Schedel is probably best known for his phantasmagorical monsters harking from the then unknown East—a dog-headed man, a one-eyed individual, a headless figure with his face in his chest, a one-legged man with an enormous foot with which he shades himself, a figure with elephantine ears extending below his hips, a figure with a long S-shaped neck with a beak in lieu of a mouth—in all fourteen such figures.

Many genealogies adorn the work. I lost count at about 150 for the number of what Schedel (1493; Schmauch 1941; Rosenberg and Grafton 2010) variously called lineages and genealogies of real and imagined personages. The most elaborate are those termed genealogies, showing vining tendrils extending from one couple to another. A visually interesting example traces the ancestry of Henry II, Holy Roman Emperor (972–1024) (see figure 2.15*B*). Regally adorned and holding a cathedral, the sainted Henry occupies the upper-left-hand corner. The genealogy starts in the bottom middle with Liudolf, the first duke and king of Saxony, and his unnamed wife. Presumably alluding to the woman giving birth, each vine emanates from the belly of the mother. In a bit of stylistic irony, the male figure below Henry II is his father, Henry II the Quarrelsome, grasping the vine that gives rise to his son and future emperor. This touch occurs in other genealogies.

Not all early trees represented specific religious or familial genealogies. One such curious tree appears in Isidore of Seville's amazing *Etymologiae: De summo bono*. Illustrated manuscript versions of this work dating from around his lifetime as well as incunabular printed versions from the late fifteenth century survive

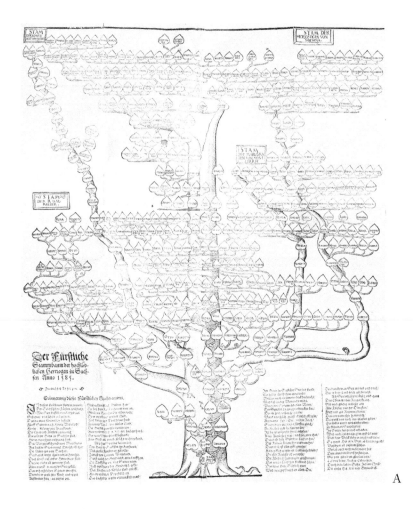

A

B

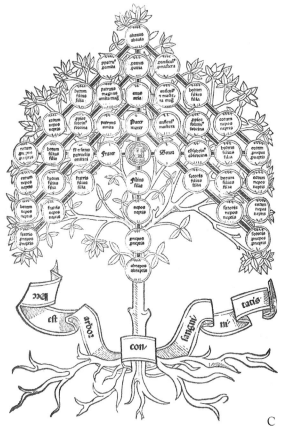

C

(Isidore 1483). Isidore (ca. 560–636) served as the archbishop of Seville until his death. He converted the Visigoths from one form of Christianity (Arianiam) to Catholicism but was not strictly sectarian in his approach to his flock. In his work, he integrated older Greek and Roman ideas with those of the barbarians into his views of Christianity and the world. He attempted the large task of writing an encyclopedia of all knowledge of the then presumed inhabited world. Although hardly cartographically instructive, he produced a now famous small circular pie diagram showing Asia, Africa, and Europe as slices.

Somewhat stylized but otherwise naturalistic, the roots, trunk, branches, and leaves of the tree in the *Etymologiae* form the underlayer in the diagram (see figure 2.15C). Upon this, superimposed circular medallions align along a series of connected thick, black lines. These lines do not follow the growth pattern of the tree but are upside down like the stemmata and trees noted earlier, yet the metaphor of a tree-like form is clear. The Latin "hec est arbor con-sangui-ni-tatis" on the ribbon above the roots means "this is the tree of shared blood"—that is, a family tree. The text in the pages preceding and following the tree figure (Isidore 2011:206–19) makes it clear that the tree does not show the ancestry of Isidore.

Isidore (2011) writes, "The family tree that legal advisers draw up concerning lineage is called a stemma [pedigree or family tree] where the degrees of relationship are spelled out—as, for example, 'this one is the son, this one is the father, this one the grandfather, this one the relative on the father's side,' and all the rest. Here are the figures for these relationships" (210). Although a bearded man (Isidore?) is depicted in a medallion in the middle of the tree, no names of relatives appear in the surrounding medallions. Rather, the medallion immediately above the bearded man reads "father mother," and ascending from there are "grandfather grandmother," "great-grandfather great-grandmother," and finally, at the very top, "great-great-grandfather great-great-grandmother." Immediately below, trailing down the trunk of the tree, are "son daughter," "grandson granddaughter," "great-grandson great-granddaughter," and finally, at the bottom, "great-great-grandson great-great-granddaughter." To the right of the medallion of the man we find "brother" and to the left "sister" medallions. Emanating from these, medallions display various familial names such as cousin, aunt, and uncle with a host of modifiers (brother's nephew, their daughter, great uncle, and so on). For Isidore, as with many ancient scholars, understanding original meanings of words was very important because such understanding yielded direct knowledge of nature. These words, which were more than arbitrary labels, instead embody the essence of the thing itself (Arnar 1990). What better way to represent this than as a tree, even if the metaphor seems upside down? Almost 1,500 years later in biological trees, the names of specific species, families, and orders came to represent members in God's creation and then, shortly thereafter, the evolutionary history of life.

FIGURE 2.15 (A) Tree tracing the history of Saxon rulers, from Lorenz Faust's *Anatomia statuae Danielis* (1585); (B) tree tracing the ancestry of Henry II, Holy Roman Emperor, from Hartmann Schedel's *Registrum huius operis libri cronicarum* (1493); (C) "Tree of Shared Blood," from Isidore of Seville's *Etymologiae* (1483).

The Sacred, Profane, and Biological Meet

With the influence of Joachim de Fiore's trees showing human history in a Judeo-Christian tradition, Isidore's even earlier tree showing the etymological origins of familial relationships, and the emergence of family genealogies figured as trees, inevitably someone would suggest that nature also be presented literally as a tree and not simply metaphorically, as the ancient Romans did in the Ara Pacis. We do not know with certainty when and how this iconography first represented the history of a life as a tree in diagrammatic form, but the German naturalist Peter Simon Pallas (1741–1811), who worked in Russia, envisioned such a diagram.

Pallas described a systematic arrangement of all organisms in the image of a tree in *Elenchus zoophytorum* (1766):

> But the system of organic bodies is best of all represented by an image of a tree which immediately from the root would lead forth out of the most simple plants and animals a double, variously contiguous animal and vegetable trunk; the first of which would proceed from mollusks to fishes, with a large side branch of insects sent out between these, hence to amphibians and at the farthest tip it would sustain the quadrupeds, but below the quadrupeds it would put forth birds as an equally large side branch. (23–24; Archibald 2009)

We must show caution in interpretation. Whereas today we see evolutionary history in such trees, almost certainly Pallas did not intend this. He was more clearly indicating plants' and animals' relative position one to another, as done in the *scala naturae* discussed in chapter 1, except now it is a tree. This appears very similar in intent to Isidore's tree of familial relationships; certainly it was accepted that such relationships resulted from consanguinity, yet there almost certainly was no intent in Pallas's musings that his tree of relationships was a result of common descent.

In 1801, thirty-five years after Pallas's suggestion, an obscure French botanist produced such a tree, a tree showing the systematic arrangement, ironically, of trees replete with leaves giving the names of tree species. The age of biological trees of life had begun, and evolutionary explanations rapidly followed, but the first half of the nineteenth century also saw other visual metaphors for nature's order that for some time lessened the impact of tree imagery.

CHAPTER THREE

Competing Visual Metaphors

Beginning in the mid-eighteenth century, European scientists struggled to keep pace with the work of classifying organisms brought back in great batches from overseas expansionary expeditions. The world's biological expanse and richness rapidly became apparent. Nevertheless, placement of this cornucopia of new species appeared mostly in dry tables and lists that only barely touched the overarching idea of organic organization let alone how it originated, except by resorting to God's creative powers.

Especially the first half of the nineteenth century saw a hodgepodge of competing ways to illustrate and organize nature's order, and as we now know, the tree won out in the end, but that was not preordained; rather, the influence of powerful individuals led to the supremacy of the tree as the prominent visual metaphor. Certainly, some illustrative metaphors were better than others for capturing nature's order—the tree, to be specific—but others were possible; some simply collapsed because of their contrived perceptions of the natural world. It nevertheless is worthwhile to sample the diversity of these perceptions.

Tables, Hierarchies, and All Manner of Other Geometries

Nothing preordained that tree imagery would triumph as the visual metaphor for nature's order. Indeed, the ladder of life, in its inaccurate simplicity and because of its historical baggage, still stubbornly persists. In the late eighteenth and earlier part of the nineteenth centuries, all manner of devices appeared in an attempt to simultaneously understand the richness and the obvious order in nature. Some of the schemes certainly wished to fathom the mind of God; increasingly, though, they stood on their own right.

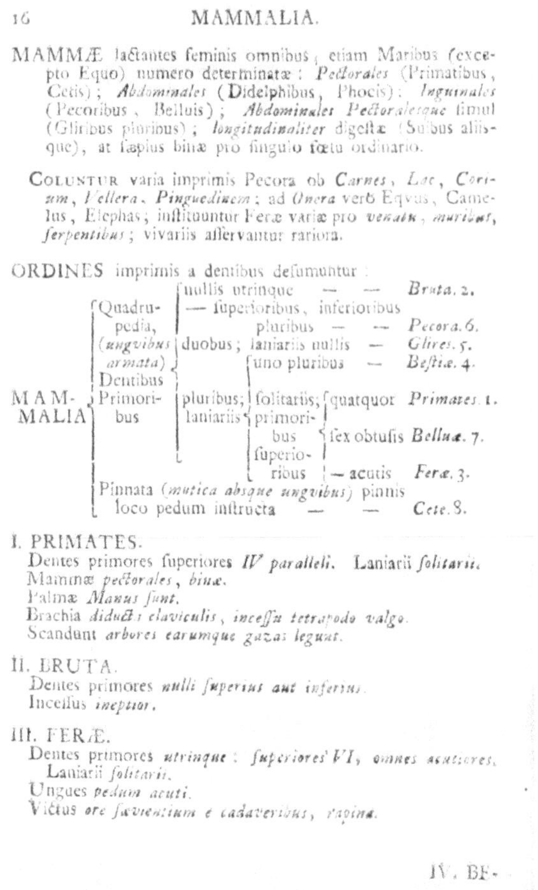

MAMMÆ lactantes feminis omnibus, etiam Maribus (excepto Equo) numero determinatæ: *Pectorales* (Primatibus, Cetis); *Abdominales* (Didelphibus, Phocis); *Inguinales* (Pecoribus, Belluis); *Abdominales Pectoralesque* simul (Gliribus pluribus); *longitudinaliter* digestæ (Subus aliisque), at sæpius binæ pro singulo fœtu ordinario.

COLUNTUR varia imprimis Pecora ob *Carnes*, *Lac*, *Corium*, *Vellera*, *Pinguedinem*; ad *Onera* verò Equus, Camelus, Elephas; instituuntur Feræ variæ pro *venatu*, *muribus*, *serpentibus*; vivariis asservantur rariora.

ORDINES imprimis a dentibus desumuntur:

nullis utrinque — — Bruta.2.
superioribus, inferioribus pluribus — — Pecora.6.
Quadrupedia (*unguibus* armata) Dentibus duobus; laniariis nullis — Glires.5.
Primoribus uno pluribus — Bestiæ.4.
pluribus; solitariis, quatuor Primates.1.
laniariis primoribus sex obtusis Belluæ.7.
superiorius — acutis Feræ.3.
Pinnata (*mutica absque unguibus*) pinnis loco pedum instructa — — Cete.8.

I. PRIMATES.
Dentes primores superiores IV paralleli. Laniarii solitarii.
Mammæ pectorales, binæ.
Palmæ Manus sunt.
Brachia diducta claviculis, incessu tetrapodo vulgo.
Scandunt arbores earumque gazas legunt.

II. BRUTA.
Dentes primores nulli superius aut inferius.
Incessus ineptior.

III. FERÆ.
Dentes primores utrinque: superiores VI, omnes acutiores,
Laniarii solitarii.
Ungues pedum acuti.
Victus ore sævientium e cadaveribus, rapina.

IV. BE-

FIGURE 3.1 Carl Linnaeus's hierarchical key of Mammalia, from *Systema naturæ* (1758).

These schemes ran from simple tables listing plant and animal names, to hierarchically organized tables and figures with progressively more exclusive groupings, to elaborate geometric shapes purporting to show some underlying mathematical principle in biology. Such schemes survive and thrive in the form of tabular classifications and dichotomous keys used by befuddled students and nature lovers alike to identify species. Who has not at least thumbed through a bird guide or looked through a key to local flowers based on their color? Some of the earlier tables and hierarchies require some comment, but they seldom rise to the level of visual impact that we associate with trees or ladders of life. I mention only the more visually appealing and scientifically important among the older examples.

Although not the first, Carl Linnaeus (1707–1778) rightfully occupies the position of most influential arranger of plants and animals, in large measure because of his binomial system of naming species whereby a plant or an animal receives both a more inclusive generic name (for example, *Felis*) and a specific or trivial name (for example, *catus*), such as the domestic cat (*Felis catus*). Linnaeus did not discover this arrangement but applied it and in so doing helped bring stability to the chaotic naming of species and higher taxa. Linnaeus left no indication of evolutionary ideas, yet his classification is possible because of evolution. With the advent of newer views of classificatory schemes and the explosion of molecular technology, a strict Linnaean system is increasingly under fire. Suffice it to note here that however well his ideas have fared, he remains the place of beginning for modern classification.

Linnaeus wrote twelve editions of his classification *Systema naturæ* beginning in 1735, but the tenth edition (1758) by agreement serves as the starting place for all later zoological classification. As a simple example of a hierarchical key, figure 3.1 comes from Linnaeus (1758). The key in the middle of the page starts with the more inclusive Mammalia on the left, providing more and more exclusive characters of posture, teeth, and so forth until the individual orders are reached on the right, with the correct page number as to where they may be found in the volume. The top of the page details characters of the mammary glands found in mammals, and the bottom lists features found in three of the orders that Linnaeus recognized. Classificatory keys and hierarchies appear relatively often by the late eighteenth century, but some are as much as two hundred years older, such as a bifurcating key from 1592 for species of hyacinths illustrated by Mark Ragan (2009).

The nineteenth century presented some of the most interesting quackery found in any century—spiritualism and phrenology are two of the best examples. Lest we think that these were fringe practices, recall that Sir Arthur Conan Doyle and Alfred Russel Wallace were spiritualists and that Queen Victoria and Prince Albert invited the notorious George Combe to perform phrenological readings of their children's heads. This is the same century in which the scientific method truly began to emerge as a way of exploring the physical world, but this also meant that rather unusual (to put it mildly) hypotheses emerged on how to visualize and organize the natural world.

In 1766, Peter Simon Pallas not only suggested trees and networks for organizing life but also mentioned polyhedrons as possible representations of nature, and in fact fifty years later a rather bewildering diversity of polyhedric representations of nature emerged (Ragan 2009). The best-known, and in England the most widely accepted, version was quinarianism. The basic thesis was that there were cycles within cycles centering on groups of five. The English ornithologist William John Swainson (1789–1855) helped popularize the system (O'Hara 1991). Swainson credited quinarianism's rise in the early nineteenth century to German-born Russian invertebrate biologist Gotthelf Fischer von Waldheim (1771–1853), who in 1805 arrayed animals in a series of contiguous circles with man at the center (Ragan 2009).

William Sharp Macleay (1792–1865) was the person most responsible for the rise of quinarianism. A Cambridge University–trained Australian amateur entomologist and son of the entomologist Alexander Macleay, William Macleay believed that he had uncovered a five-part division of all animals that he termed the Acrita, Radiata, Annulosa, Vertebrata, and Mollusca (figure 3.2). Recall that Georges Cuvier, whom Macleay met, had recognized four branches of animals. But unlike Cuvier's system, Macleay claimed that each of the five groups was divisible into five lesser groups, such as Pisces, Amphibia, Reptilia, Aves, and Mammalia within Vertebrata. Macleay was thus claiming something innate about divisions of five in the natural world.

On the same page that his diagram of the five groups appears, Macleay (1819) writes, in reference to various groups of animals: "The foregoing observations I am well aware must be far from accurate; but they are sufficient to prove that there are five great circular groups in the animal kingdom which possess each a peculiar structure, and that these, when connected by means of five smaller osculant groups, compose the whole province of Zoology" (318). By "osculant groups," he meant intermediary forms. For example, Cephalopoda linked Vertebrata and Mollusca, and the platypus linked mammals and birds (Oldroyd 2001). If five members of some grouping could not be found, it meant that the missing members remained to be recovered, such as the case for the asterisks in Mollusca in figure 3.2.

The quinarian system represented an interesting, if misguided, attempt to recover and visually represent the relationships of groups of species, and it was ahead of its time in trying to ascertain what might be missing in terms of undiscovered groups of species. The latter idea resurfaced in more recent time with a

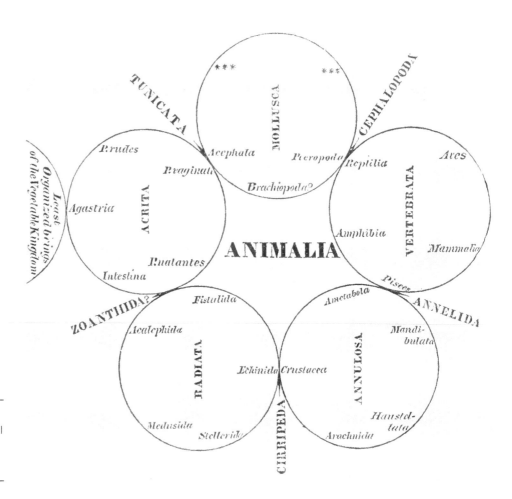

FIGURE 3.2 William Macleay's
quinarian, or five-part, division of all
animals, from *Horae Entomologicae*
(1819).

"ghost lineage," which predicts the presence of its nearest relative in the fossil record (Norell 1992) by comparing what is known of the earliest fossil record of one lineage with what is not known of its nearest related lineage. Although a hypothesis, ghost lineages provide a basis for possible discovery, whereas such a premise in quinarianism does not and never did have any basis in biology. Thus, while testable, the idea of a five-part overarching basis for nature was a contrived hypothesis even in its day and soon collapsed of its own internal inconsistencies.

The quinarian system was adopted by various researchers, mostly British, in the first half of the nineteenth century, but it became increasingly cumbersome and complex, going from five to seven and then to ten cycles, and as Ragan (2009) notes, "these systems began a terminal slide into disfavor" (13). Nonetheless, the system appeared in various well-known, mostly popular texts in the first half of the nineteenth century, notably as a chapter in the evolutionist Robert Chambers's (1802–1871) anonymously published *Vestiges of the Natural History of Creation* (1844) and in the creationist Hugh Miller's (1802–1856) *Testimony of the Rocks* (1857). It even survived in a few places after the publication of *On the Origin of Species* in 1859; William Hincks, a professor of natural history at the University of Toronto who had been chosen for his position over a bright, young Thomas Henry Huxley, was teaching the quinarian system as late as 1870 (Coggon 2002).

Pallas's *Elenchus zoophytorum* (1766), which suggests that life be arranged in a tree-like form, traditionally marks the point in time when European savants first put forward that idea. Pallas never produced such a tree figure, at least of which we are aware, but sixty-three years later a figure produced by Carl Edward von Eichwald (1795–1876) in *Zoologia specialis* (1829) purports to depict Pallas's tree (Ragan 2009). Ragan claims that according to a Russian source, S. R. Mikulinskii (1972), Eichwald's diagram shows Pallas's tree. In fact, although Mikulinskii notes the works both of Pallas and of Eichwald in the same paragraph (Chesnova 1972:344), she presents no evidence that Eichwald intended his tree to be a visual version of Pallas's ideas. Nonetheless, Ragan's close reading of Pallas shows the considerable similarity of Pallas's description of a tree and Eichwald's figure of one. According to Ragan, Pallas describes the trunk of his tree as a series of neighboring genera (or lineages) closely appressed one to another and with twigs thrusting from the trunk—exactly what Eichwald presents (figure 3.3). Eichwald (1829) does give credit to others for earlier uses of a tree metaphor when he writes (in Latin), "e) Long ago other authors constructed a nearly similar animal tree of life; foremost among whom must be numbered my very good friend F. S. Leuckart (v Zoologische Bruchstiicke, Helmstädt, Heft I., 1819, and the same Versuch Dying naturgemässen Eintheilung der Helminthen, Heidelberg, 1828), or A. F. Schweigger (v Naturgeschichte der skelettlosen, ungegliederten Thiere, p. 81)" (44). A search of these works uncovered only bracketing classifications but no trees similar to that of Eichwald.

Eichwald's figure evokes a rather brooding feel, a darker Blakean view of the creation of life, showing what appears to be separate trunks closely joined into one, just as Pallas described. Roman numerals surmount eight of the larger branches, and Eichwald refers different groups of animals to each large branch. He titles this figure and the four and a half pages that follow "The Tree of Animal Life." Eichwald (1829) describes in the first few sentences how "the first rudiments of animal life [come] from the mass of chaotic organic material" that we find in the ever-present warm water and repeatedly form from a "mucous membrane of globules of various primitive forms" (41). Some of Eichwald's language suggests biological change, as in several places where he uses variations of the Latin word *evolutio*, but whether he meant anything like what we call biological evolution or only the older, preformationist sense of unrolling, as of an organism during its lifetime, remains murky. Still, Eichwald (1829:43) tempts us when we read that the highest branches of a tree, which overshadow the lower vertebrates, gradually evolve in a straight line, intricately making it up to the ascending human race, whom he then compares to spring flowers on the tree, and further that each step in the evolution of animals follows different ways, but then again he may simply mean evolution in the sense of unrolling (development) of an individual organism.

Whereas we may be tempted to read some evolutionary thoughts into Eichwald's writings, the same cannot be said for Pallas. Nevertheless, Pallas not only argued for life to be represented as a tree but most specifically stated that life should not be connected in a series in a scale or ladder (Ragan 2009). Even though

FIGURE 3.3 Carl Edward von Eichwald's tree of continuous change, from *Zoologia specialis* (1829).

Pallas rejected the *scala naturae*, following Ragan's interpretation, both Pallas's and Eichwald's ideas appear as a hybrid of a *scala naturae* and a tree that Ragan colorfully describes as a bunch of asparagus shoots rather than a tree.

Certainly Eichwald's comments and figure, whether intended or not, suggest repeated spontaneous generations followed by change in closely aligned lineages, an idea that, as we will see shortly, was earlier advocated by Jean-Baptiste Lamarck. Although it cannot be said for Eichwald, nothing of evolution imbues the ideas of Pallas, yet it seems natural for us today to read such ideas into these early works. There is little doubt that before Pallas, others may have suggested similar ideas; they simply do not survive or remain unrecognized.

The first known tree presenting biological (but not evolutionary) relationships appeared in a paper written by an obscure French botanist, Augustin Augier (1801) from Lyon (figure 3.4). This work, however, did not go unreported at the time of its publication, at least by its publisher. In 1802, the inaugural annual issue of a catalog of books of seven book publishers (Fleischer 1802), including its Paris distributor Levrault frères, listed Augier's work under its botanical section as item 59. Augier's tree even merits a quaint description: "with botanical tree of large grape leaves, with its explanation" (quoted in Stevens 1983:267). As far as determinable, Augier's tree influenced no later tree motifs and probably would not be known except for its fortuitous recognition and publication in 1983 by botanist Peter F. Stevens. This said, the publication in 1801 of such a tree shows that this motif was beginning to infiltrate scientists' views on how to show nature's order. According to Stevens, Augier was not a full-time botanist, certainly not a rarity at the time, but relied on better-known botanists of his era for his information. As Stevens's translation shows, Augier specifies a creator rather than any acceptance of evolution to explain his tree.

Augier's tree is reminiscent of Joachim of Fiore's much older trees, but instead of religious personages gracing the medallions on Fiore's tree, Augier provides medallions (the ersatz "grape leaves") showing what he terms classes and tribes of plants, at least as recognized at that time and by him. At the base is the tribe "Cryptogames," including fungi, mosses, algae, and ferns. Immediately above occurs the "Phanerogames," including what we would today refer to as angiospermous and gymnospermous plants. Each of the numerous terminal leaves bears smaller groups of plants. The stars scattered around the tree denote families that show the relationship of analogy (Stevens 1983), meaning similarities not the result of being close relatives.

Even though Augier apparently rejected the notion of a *scala naturae*, his tree retains a *scala naturae* aspect in that he places what he perceives as more primitive plants nearer the base with more "perfect" forms set higher on the main trunk. Stevens's (1983) translation shows this intent: "This method starts with the least perfect plants and by gradation leads to the more perfect, as one can convince oneself when reading the exposition of the method and the explanation of the botanic tree" (205). This dual representation continues well into the nineteenth century and beyond, even after evolution triumphed as the basis for such a tree-like form.

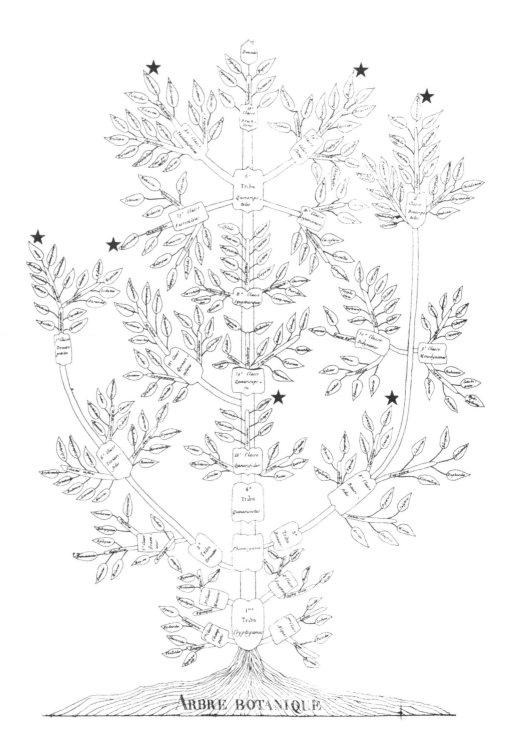

FIGURE 3.4 Augustin Augier's "Tree of Trees," from *Essai d'une nouvelle Classification des Végétaux* (1801). (Reproduced by permission of *Taxon*)

The Evolution of an Evolutionist and His Changing Visual Metaphors

Within a short eight years after the publication of Augier's (1801) rather picturesque tree, another, much less arborescent, inverted, but nonetheless branching diagram appeared, but for the first time evolution underpinned its form. Augier's full-blown tree motif provides a stark contrast to Lamarck's (1809) hesitant figure formed of widely spaced dots in a simple inverted bifurcating diagram (figure 3.5).

As far as is known, Jean-Baptiste Lamarck (1744–1829) knew nothing of Augier's tree, which seems somewhat odd as Lamarck began his career as a botanist who spoke and wrote French, as did Augier. By the time Lamarck published this figure in 1809, his views on how the evolutionary history of life unfolded had themselves evolved, somewhat remarkable for a scientist of almost sixty-five years of age.

At the age of only thirty-four, Lamarck fairly burst on the biological scene with the publication of his over 1,800-page, three-volume botanical work *Flore françoise* (1778), followed a year later by his rapid election to the French Academy of Sciences. He certainly received help from luminaries at the Jardin du Roi (Jardin des Plantes after the French Revolution) in Paris, which he readily admits in the introductory parts of this work, but he nevertheless begins here to show his own mind. The volumes present a series of mostly dichotomous keys and classifications of plants, quite different from the approach of Linnaeus (Packard 1901).

In the preliminaries of volume 1, Lamarck posits three problems dealing with how the succession of plants should begin and proceed, what rules provide these arrangements, and how to order such a succession for which there are no lines of demarcation to divide groups. Lamarck's choice to deal with these problems, especially the last, comes in the form of a four-page table, which he says samples the natural order using exemplars. Figure 3.6 is a composite of the four pages and is read from the top-left down and then from top-right down. I am not aware of why Lamarck chose to progress these tables from top to bottom, but, as we will see, he does this for some later figures but not others, even though form and intent change.

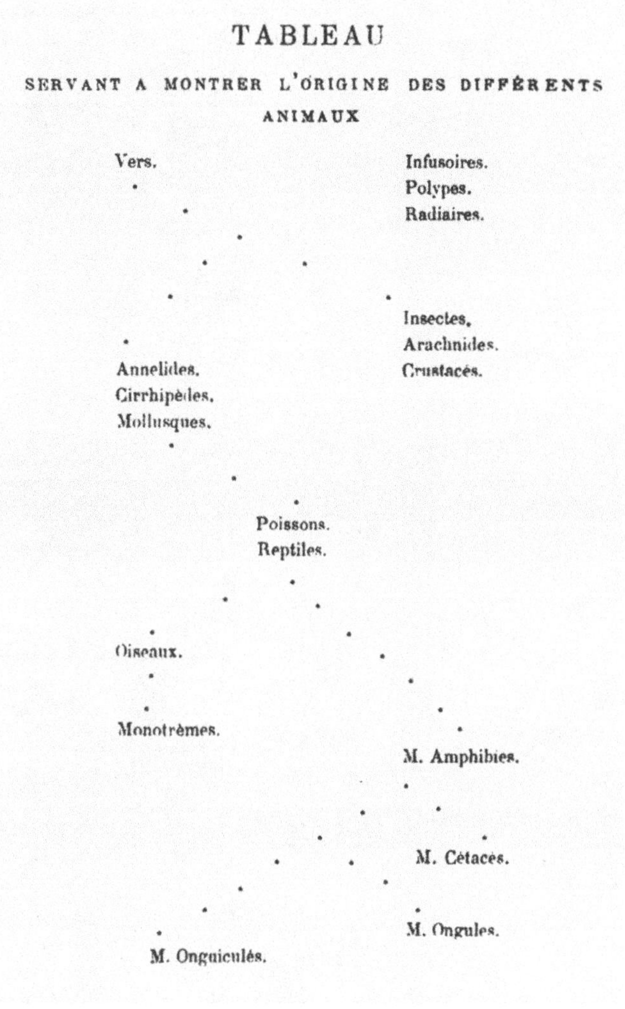

The column headings from left to right indicate a series of closely related genera, the middle column notes specific saliences or characters derived as the result of notable affinities, and the right column refers to more distant, general relationships indicating the gradual perfection of organs. (In the middle column, I provide English translations of the major groups recognized by Lamarck.) Relative to modern classifications, the accuracy of Lamarck's table varies; for example, fungi is now regarded as its own kingdom, separate from plants. Although read from top to bottom, the table represents a very finely grained *scala naturae*, much more so than that of Charles Bonnet (see figure 1.4). Further, unlike for Bonnet, we see no horizontal divisions in Lamarck's tables. It may not be so much that Bonnet wished to show such divisions, but it is clear according to Stevens (1994) that Lamarck intended a continuous *scala naturae*, with the result that demarcations or limits of higher taxa were arbitrary. As Stevens further notes, Lamarck thought that naturalists erroneously confused reconstructing the order of nature

TABLEAU

SERVANT A MONTRER L'ORIGINE DES DIFFÉRENTS ANIMAUX

Vers.

Infusoires.
Polypes.
Radiaires.

Insectes.
Arachnides.
Crustacés.

Annelides.
Cirrhipèdes.
Mollusques.

Poissons.
Reptiles.

Oiseaux.

Monotrèmes.

M. Amphibies.

M. Cétacés.

M. Ongulés.

M. Onguiculés.

FIGURE 3.5 The first-known evolutionarily based tree, "serving to show the origin of the different animals," from Jean-Baptiste Lamarck's *Philosophie zoologique* (1809).

SÉRIE GÉNÉRALE des genres rapprochés en raison de leurs rapports.	SAILLIES PARTICULIÈRES formées par certaines affinités remarquables.	RAPPORTS GÉNÉRAUX & éloignés, indiquant la perfection graduée des organes.
Agaricus T. Boletus. Fungus. Hydnum. Phallus. Elvela. Clathrus. Peziza. Lycoperdon. Clavaria. Mucor. Byssus. Conferva. Ulva. Tremella. Fucus. Lichen. Targionia. Anthoceros. Riccia. Blasia. Marchantia. Jungermannia. Buxbaunia.	**Fungi** *Champignons.* Substance spongieuse, lamellée ou poreuse, & qui, sous diverses formes, s'étend en hauteur ou est très-ramassée. **Algae** *Algues.* Substance aplatie, membraneuse, & qui, sous diverses ramifications, s'étend en longueur, & produit des cupules floriformes.	Fructification absolument inconnue & insensible.

Suite DE LA SÉRIE formée par le rapprochement des genres.	SAILLIES PARTICULIÈRES formées par certaines affinités remarquables.	RAPPORTS GÉNÉRAUX & éloignés, indiquant la perfection graduée des organes.
Zamia. Cycas. Chamærops. Sanbal. Borassus. Corypha. Cocos. Elate. Areca. Caryota. Elais. Phœnix. Calamus. Flagellaria. Oryza. Zizania. Pharus. Olyra. Paspalum. Antoxanthum. Alopecurus. Phleum. Phalaris. Panicum. Milium. Stipa. Agrostis.	**Palms** *Palmiers.* Feuilles ramassées en faisceau au sommet de la tige qui est simple. Fleurs paniculées & enfermées dans un spathe.	Fructification sensible & très-distincte; étamines de deux à six; semences ordinairement nues & solitaires.

Suite DE LA SÉRIE formée par le rapprochement des genres.	SAILLIES PARTICULIÈRES formées par certaines affinités remarquables.	RAPPORTS GÉNÉRAUX & éloignés, indiquant la perfection graduée des organes.
Hypnum. Brium. Mnium. Polytrichum. Splachnum. Fontinalis. Porella. Phascum. Sphagnum. Lycopodium. Equisetum. Isoetes. Pilularia. Marsilea. Ophioglossum. Osmunda. Onoclea. Pteris. Asplenium. Trichomanes. Blechnum. Hemionitis. Lonchitis. Adiantum. Acrosticum. Polypodium.	**Mosses** *Mousses.* Feuilles nombreuses, & disposées en gazon, ou embriquées autour des tiges qui produisent des urnes anthériformes. **Ferns** *Fougères.* Feuilles toutes radicales, roulées en crosse avant leur développement, & chargées de poussière séminiforme.	Fructification sensible, mais indistincte ou peu connue.

Suite DE LA SÉRIE formée par le rapprochement des genres.	SAILLIES PARTICULIÈRES formées par certaines affinités remarquables.	RAPPORTS GÉNÉRAUX & éloignés, indiquant la perfection graduée des organes.
Aira. Melica. Poa. Briza. Uniola. Dactylis. Festuca. Bromus. Avena. Holecus. Andropogon. Arundo. Lagurus. Cynosurus. Hordeum. Secale. Triticum. Elymus. Lolium. Nardus. Ægilops. Cenchrus. Carex. Eriophorum. Scirpus. Cyperus, &c.	**Grasses** *Graminées.* Feuilles simples, alongées & engainées à leur base. Fleurs enfermées dans des paillettes.	

FIGURE 3.6 Composite of Lamarck's four tables showing the progression of plants (and fungi), read from the top-left down and then from the top-right down, from *Flore françoise* (1778).

with assigning names to organisms. Although biologists now know that evolutionary change drives the system, they still haggle over how phylogenetic reconstruction should be reflected in the names applied to species and higher taxa. Also, Lamarck's views on his system did not remain static, for in 1786 he reversed and in part reorganized the whole sequence for some of the plants (Stevens 1994). These and later changes in his scientific views characterize Lamarck as a forward-looking biologist. Although the remainder of the three volumes of *Flore françoise* carry on through the angiosperms, or flowering plants, of France, for some reason Lamarck stopped with the grasses and relatives or monocotyledonous plants (flower-bearing plants with a single rather double embryo in the seed) in his four *scala naturae* tables in volume 1. Instead, near the beginning of volume 2, following the title page and four pages of advertisements, Lamarck provides a foldout of his table of principal divisions of the plants he analyzed, not unlike the much simpler table from Linnaeus (see figure 3.1). Lamarck's table includes bracketed names of major plant groups and their characteristics, leading to smaller, more inclusive groups of plants and finally numbers indicating where they might be found in the volumes. Such tables, which are neither trees nor ladders, became a common method by at least the eighteenth century of providing a convenient way to show an author's classificatory scheme. Similar schemes continue to this day.

Within the time frame of the publication of *Flore françoise*, the realization emerged that a purely linear intergradation of species in the *scala naturae* did not explain the great breadth of nature. Lamarck began his career within this framework, but it soon began to change, and he with it. We cannot be sure, but he probably began to shift his ideas about the immutability of species slightly before he started viewing the pattern of evolutionary change in any sort of bifurcating manner. There appear to be no clear statements of his ideas on the topic within his botanical works in the late eighteenth century, but by 1802, after he began to work on invertebrates, his views clearly reflect an evolutionary bent. In the appendix to *Recherches sur l'organisation des corps vivans* (1802), he writes: "I have for a long time thought that species were constant in nature, and that they were constituted by the individuals which belong to each of them. I am now convinced that I was in error in this respect, and that in reality only individuals exist in nature" (141; Packard 1901). In the next few pages in the same work, after Lamarck discusses how the changes in the physical earth relate to the biological realm as well as chastising himself and other naturalists for their blindness about the fixity of species, he states: "All changes that each living body has experienced as a result of changes of circumstances that influenced his being, will no doubt spread. . . . But as new changes necessarily continue to operate, regardless of slowness, not only will it always form new species, new genera, and even new orders" (143). Lamarck could not be more explicit in his argument for evolutionary change causing speciation and the origin of higher taxa such as orders.

In writings of Lamarck from 1801, Packard (1901) believed that he had found indirect evidence that Lamarck lectured in his courses about the possibility of evolutionary change by the 1790s. Whatever the case may be, by the earliest years of the nineteenth century, Lamarck was beginning to write about evolutionary change, which also changed his views from one of straight-line order in nature to some sort of bifurcations in nature's order.

What may be the earliest figure by Lamarck that hints at his changing views on the order of nature, if not evolution, comes in his *Encyclopédie méthodique* (1786:33). In the figure, he presents what he calls his table of living organic beings—a *scala naturae*—but of a different kind. Instead of a single line of plants and animals, on the left is the progression of animals and on the right the progression of plants (figure 3.7). Interestingly, unlike Lamarck's earlier attempt for plants, this diagram opens upward and the two branches diverge. If this were not enough, on the very next page, Lamarck presents us with another *scala naturae*, but this time

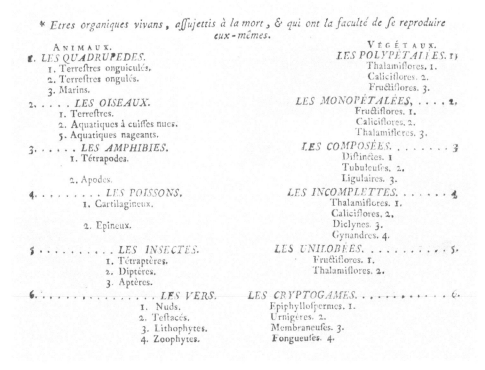

FIGURE 3.7 Lamarck's "table of living organic beings"—"Living organic beings, subject to death, and who have the ability to reproduce themselves"—from *Encyclopédie méthodique* (1786). (The English translation is by the author.)

of inorganic beings, beginning at the base with crystalline rock and diverging left upward toward animals and on the right toward plants (figure 3.8). He makes the point that in each table, these form different successions, but successions nonetheless. Lamarck has gone beyond the Bonnet-style ladder of life, or even his own earlier work, now seeing in both the inorganic and organic worlds that divergences exist, and not simply ladders.

These earlier figures along with Lamarck's comments bring his famous 1809 bifurcating diagram—the earliest known evolutionary tree—into clearer focus. He began with a *scala naturae* (for plants) in 1778 (see figure 3.6), changed to a bifurcating *scala naturae* of plants and animals diverging one from another in 1786 (see figure 3.7), as well as even more divisions of inorganic matter set below plant and animal life (see figure 3.8), and finally arrived at his bifurcating diagram in 1809 (see figure 3.5). We cannot but notice that he first uses a table opening downward, then one opening upward, and finally one downward again in 1809, but it seems no more than his changing graphic preference. Lamarck's diagram can fairly be called the first evolutionary tree of life because it marries a branching diagram with evolution as the mechanism creating the branching (Archibald 2009). In his discussion preceding the diagram, Lamarck (1809) hypothesizes that the loss of the hind limbs and pelvis in cetaceans and the similar trend in seals are a result of disuse, one of his themes for the cause of evolution:

If it is considered that, in the seals where the pelvis still exists, this pelvis is impoverished, narrowed and without hip projections; it will be felt that the poor use of the posterior feet of these animals must be the cause, and that if this use entirely ceased, the hind feet and even the pelvis could at the end disappear. . . . The following chart will be able to facilitate the intelligence of what I have just exposed. It will be seen there that, in my opinion, the animal scale starts at least with two particular branches, and that, in the course of its extent, some branches appear to finish in certain places. (462)

Lamarck's views changed over time, which results in some confusion as to his intentions at any given time. Earlier he represented the organization of life (actually plants) in a nonevolutionary, infinitely graduated *scala naturae*; later, as some branching *scala naturae* during the time he may have first contemplated evolution as a cause; and finally, as a tree-like, bifurcating figure that has been read both as an evolutionary *scala naturae* and as a figure showing descent. The dual aspects of Lamarck's theory of evolution—the inherent tendency of matter to develop increasing perfection and the adaptive power of the environment acting on animals through their needs—led Lamarck to his branching diagram (Appel 1980) but does not tell the intent of the 1809 diagram. Similarly, in his essay included in an English translation of *Philosophie zoologique*, R. W. Burkhardt (1980) addressed Lamarck's view of the pattern of evolution. Burkhardt noted that by 1802, Lamarck indicated that because of environmental influences, animal species could not be arranged linearly but formed "lateral bifurcations" (xxiv), and further that by 1815, Lamarck viewed a single line of increasing complexity as untenable (xxxiii).

FIGURE 3.8 Lamarck's "table of inorganic beings"—"Inorganic beings, lifeless, and produced by successive alterations of compound substances that were part of living beings"—from *Encyclopédie méthodique* (1786). (The English translation is by the author.)

Evolution unquestionably provides the means of change shown in Lamarck's bifurcating diagram. We must, however, be cautious in interpreting his full intent. As Peter Bowler (1989) clearly articulates, we might assume that Lamarck intended to show in such a diagram the evolution of species from a common ancestor whose descendants adapted and transformed with environmental changes. In fact, Lamarck likely never accepted the idea that living forms descended from a common ancestor. Rather, his view was one of a continually changing *scala naturae* in which organisms alive today progressed to their present stage separately, probably not unlike what Eichwald (1829) represented in his tree (see figure 3.3). Organisms at different levels of complexity arose from separate events of spontaneous generation at different times shown in his evolutionary diagram. The earlier quote regarding Lamarck's (1809) bifurcating diagram that "the animal scale starts at least with two particular branches" and that "some branches appear to finish in certain places" (462) makes more sense when viewed in this context.

For Lamarck, what we now call evolution was an ongoing process of multiple spontaneous generations with repeating bifurcations, yielding a tree that shows lines tracking through repeated divergences of one form after another as complexity increased.

Lamarck was not alone in presenting such diagrams. According to Pascal Tassy (2011), Charles-Hélion de Barbançois (1760–1822) reproduced Lamarck's tree in 1816, adding greater detail. It was a giant, poster-size, difficult to reproduce figure showing the history from single-celled creatures to monkeys. Tassy notes that the French term *filiation* was used for the first time, in this context as an unequivocal phylogenetic concept. Although quite fanciful, Barbançois's tree shows extant groups giving rise to other groups. In 1841, the biologist William Benjamin Carpenter (1813–1885) used a visual metaphor similar to that of Lamarck's (1809) tree. Unlike either Lamarck's simple bifurcating diagram or Barbançois's much more elaborate tree, both of which tracked evolutionary change, Carpenter's diagram tracked differences in timing of embryological development (figure 3.9*A*).

Evolving Ontogenetic Diagrams into Phylogenetic Trees

Carpenter was an adept systematizer of the newer biological sciences of the mid-nineteenth century (Desmond 1989). This is well represented by his book *Principles of General and Comparative Physiology* (1839). Although the volume did include physiology, it dealt in a very major way with what we today call comparative anatomy and development. It covered plants and both invertebrates and vertebrates. Without denying that there may be four great separate Cuvierian animal *embranchements*, or branches of life, Carpenter found strong ties between these groups, in part through their developmental biology.

The work of embryologists such as Karl Ernst von Baer (1792–1876) greatly impressed Carpenter as well as Carpenter's contemporaries with the argument that during development general characters across various groups of animals appear before the more specific characters of the group appear. He also argued that the embryos of so-called higher animals never resemble the adult forms of so-called

lower animals. Both a fish and a mammal have outpock-
etings in the throat region in earlier stages of embryonic development that go on to develop into true gills
in fishes but not in mammals. The mammalian embryo
does not replicate adult fish gills in its development.
Notably, Carpenter did come to support in a somewhat limited fashion Charles Darwin's (1859) vision of
evolution, and although von Baer finally accepted that
evolution was likely, he did not accept Darwin's natural
selection (S. J. Holmes 1947).

Carpenter's diagram (see figure 3.9*A*) does not
appear in the first edition of *Principles* in 1839 but only in
the second edition in 1841 and then not in later editions.
I have not found any compelling reasons why this is so.
In the preface to the second edition, Carpenter (1841)
lists "important additions and alternations" (xi–xii) to
numbered paragraphs in the first edition, including the paragraph altered to give a
description of figure 3.9*A*. The first part of paragraph 244 in the second edition corresponds to the entirety of paragraph 203 in the first edition (172), but to it has been
added the quite small figure as well as discussion of the figure. This new section of
paragraph 244 providing the figure discussion in part reads:

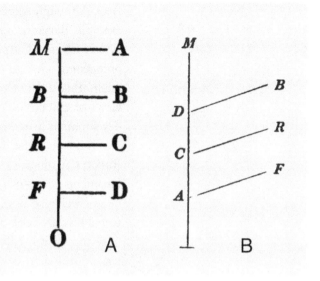

FIGURE 3.9 (A) William Carpenter's diagram tracking differences
in timing of embryological development, from *Principles of General and
Comparative Physiology* (1841);
(B) Robert Chambers's very similarly drawn diagram, indicating how
embryological changes can be interpreted as an evolutionary history,
from *Vestiges of the Natural History
of Creation* (1844).

> It is to be remembered that every Animal must pass through *some* change, in
> the progress of its development from its embryonic to its adult condition; and
> the correspondence is much closer between the embryonic Fish and the fœtal
> Bird or Mammal, than between these and the adult Fish. . . . The new here
> stated may perhaps receive further elucidation from a simple diagram. Let the
> vertical line represent the progressive change of type observed in the devel-
> opment of the fœtus, commencing from below. The fœtus of the Fish only
> advances to the stage *F*; but it then undergoes a certain change in its progress
> towards maturity, which is represented by the horizontal line *F*D. The fœtus
> of the Reptile passes through the condition which is characteristic of the
> *fœtal* Fish; and then, stopping short at the grade *R*, it changes to the perfect
> Reptile. The same principle applies to Birds and Mammalia; so that A, B, and
> C,—the *adult* conditions of the higher groups,—are seen to be very different
> from the *fœtal*, and still more from the *adult*, forms of the lower. (196–97)

This passage clearly reiterates von Baer's ideas of general development, yet no
mention of von Baer appears, and there is no clear indication of evolutionary
change. In the second edition of *Principles* (1841), Carpenter mentions von Baer
only twice in passing, yet Carpenter recognized von Baer's ideas in the first edition (1839:170), so why not in the second, especially in the context of his figure?
As it turns out, this was an unintended oversight. By the fourth edition (1854
[American edition viewed]), Carpenter frequently cites von Baer. In an extended
footnote in this edition, Carpenter gives full credit and praise to von Baer for his
"great developmental law" (126), noting the use of figures similar to those of von

Baer. Although sometimes presented as some sort of rudimentary phylogeny (for example, Barsanti 1992), and as much as we might wish to breathe phylogeny (and evolution) into Carpenter's branching diagram, it deals solely with von Baerian ontogenetic change from the more generalized embryo to the more specialized adult. This did not mean he accepted Richard Owen's more idealized Platonic views of embryological morphotypes; rather, Carpenter's views of embryological development firmly resided with the materialist perspective (Desmond 1989).

This is not the case for a similar branching diagram, which first appeared in Robert Chambers's *Vestiges of the Natural History of Creation* (1844) (see figure 3.9*B*). Chambers (1802–1871) was a successful Scottish publisher, not a scientist, although he was well read in geology. Because of the controversial nature of *Vestiges*, he chose to publish his book anonymously. His authorship was suspected but not revealed until the twelfth edition of the book (1884), well after his death in 1871. His book was a great success with the public; nevertheless, biologists of the time generally discounted it as unscientific (Secord 2000).

As Stephen Jay Gould notes in *Ontogeny and Phylogeny* (1977), Chambers's diagram comes almost straight from Carpenter without attribution. Nor is there attribution of the embryological ideas of von Baer. Chambers's diagram varies only slightly from Carpenter's, yet his description transforms it from one of von Baerian embryonic change, as described in Carpenter's text, into an explication of evolutionary change. The allusion to there being smaller "ramifications" completes the tree analogy, even if Chambers (1844) is not explicit in calling it such:

> It has been seen that, in the reproduction of the higher animals, the new being passes through stages in which it is successively fish-like and reptile-like. But the resemblance is not to the adult fish or adult reptile, but to the fish and reptile at a certain point in their foetal progress; this holds true with regard to the vascular, nervous, and other systems alike. It may be illustrated by a simple diagram. The foetus of all the four classes may be supposed to advance in an identical condition to the point A. The fish there diverges and passes along a line apart, and peculiar to itself, to its mature state at F. The reptile, bird, and mammal, go on together to C, where the reptile diverges in like manner, and advances by itself to R. The bird diverges at D, and goes on to B. The mammal then goes forward in a straight line to the highest point of organization at M. This diagram shews only the main ramifications; but the reader must suppose minor ones, representing the subordinate differences of orders, tribes, families, genera &c., if he wishes to extend his views to the whole varieties of being in the animal kingdom. Limiting ourselves at present to the outline afforded by this diagram, it is apparent that the only thing required for an advance from one type to another in the generative process is that, for example, the fish embryo should not diverge at A, but go on to C before it diverges, in which case the progeny will be, not a fish but a reptile. To protract the *straightforward part of the gestation over a small space*—and from species to species that space would be small indeed—is all that is necessary. (212–13)

The branching diagrams in Carpenter's second edition of *Principles* (1841) and Chambers's first edition of *Vestiges* (1844) are nearly identical. Yet essentially the same iconography has been transformed from showing developmental changes occurring from embryo to adult within four major kinds of vertebrates into an argument of how these embryological changes can be interpreted as an evolutionary history. Interestingly, in this instance the tree of life iconography has here been commandeered for a new *scientific* purpose from a previous one rather than from a nonscientific source—the scientific method at its best—testable and falsifiable.

God's Trees

With hindsight, we might think that after Lamarck not only would tree-like diagrams have become prevalent in representing nature's order but evolution would have become the process to explain these branchings. This was not to be. It was thirty-five years before Chambers anonymously published his small figure expounding on developmental change driving evolution, and it would be another fifteen years until Darwin published his one and only bifurcating foldout diagram in *On the Origin of Species* (1859), which unequivocally indicated that evolution or transmutation was the process driving the bifurcating. A few tree-like diagrams that toyed with the idea of evolution appeared in the years between Lamarck in 1809 and Darwin in 1859, and these will be discussed, but during most of this time the trees that appeared claimed God's divine hand rather than evolution as the process.

Today when we view a tree diagram replete with names or representations of plants or animals, we visualize "evolutionary history"; why would we not do so? Figure 3.10 shows a nice tree-like figure, and the French title, translated as

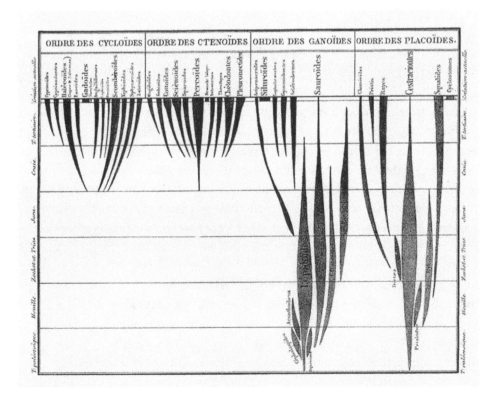

FIGURE 3.10 Louis Agassiz's geologically based, nonevolutionary "Genealogy of the Class Fishes," from *Recherches sur les poissons fossiles* (1844).

"Genealogy of the Class Fishes," surely indicates or at least implies evolution. That the classification is out-of-date is clear from the four orders listed across the top naming kinds of fish scales, a feature that certainly shows an evolutionary stamp but no longer forms the basis for fish classification. The horizontal lines demarcate various geologic times listed on either side, starting in the Paleozoic at the base, again in French, and ending at the top with "Creation actually," which may be translated colloquially as "present-day appearance or formation." Thus we seem to have not only the evolutionary history of fish but its geologic framework as well. The relative thicknesses of each line of fishes indicate relative abundances at any given time in the geologic past, known today as spindle diagrams because of their spindle-like shape. Although outdated, the diagram presents a hypothesis of relationships.

This diagram, by Swiss-born Harvard paleontologist Louis Agassiz (1807–1873), appeared in his *Recherches sur les poissons fossils* before he first visited and then moved to the United States in 1846. Although sometimes presented as being published in 1833 (for example, Voss 2010), the diagram did not appear until Agassiz's final volume of the work, published in 1844 (Brown 1890; Archibald 2009). In describing the figure, Agassiz writes:

> the family trees on the trunk of which will be registered the oldest kinds, while the branches will bear the names of the more recent types . . . the time of appearance of each group [is shown] by means of horizontal lines which will cut the branches at various heights . . . [the] intensity of the development of each family to each time, [is shown] by giving to the various branches of each kind a degree thickness in connection with the importance of the role . . . in each geological formation . . . [the diagram] represents the history of the development of the class of fish through all the geological formations and which expresses at the same time the degrees of affinity that between them various families have . . . [and] finally the convergence of all these vertical lines indicates to the affinity families with the principal stock of each kind. (170)

At the bottom of the page, Agassiz adds: "I however did not bind the side branches to the principal trunks because I have the conviction that they do not descend the ones from the others by way of direct procreation or successive transformation, but that they are materially independent one from the other, though forming integral part of a systematic unit, whose connection can be sought only in the creative intelligence of its author" (170). Thus, as odd as it might seem with our modern perceptions and biases, Agassiz did not intend to represent evolution in this diagram. Published in 1844, it appeared fifteen years before the publication of Darwin's *On the Origin of Species*.

Agassiz's (1844, not 1833) diagram of fish relationships, at first glance seemingly the earliest evolutionary paleontological tree of life, neither represents the oldest such paleontological tree of life nor shows evolutionary relationships. Today it is the best known of such diagrams, but in its day it certainly was not. Agassiz's fish diagram appeared in print only once in his lifetime. Edward Hitchcock (1793–1864)

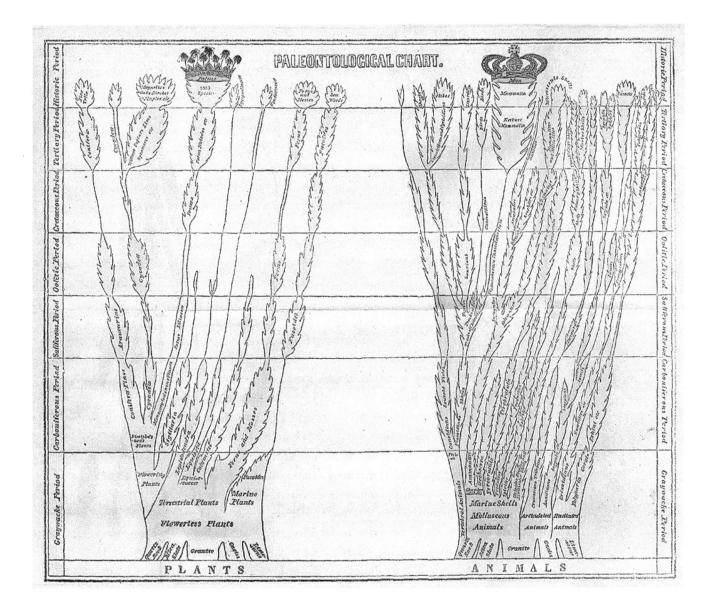

had published two trees of similar intent four years earlier, in 1840. Although little known today, in its time his "Paleontological Chart" (figure 3.11) would certainly have been better known than Agassiz's fish diagram, if for no other reason than Hitchcock's trees of life were published in his popular *Elementary Geology* in thirty editions from 1840 to 1856 (or possibly 1859), with another edition, lacking the chart, written with his son Charles between 1860 and 1866.

For those who might know of Hitchcock today, it would be for his tenure as president of the small liberal arts Amherst College (1845–1854) and for his description of dinosaur footprints in the Connecticut River valley. Hitchcock held fast to the notion that these footprints were those of giant birds not belonging to Richard Owen's new-fangled dinosaurs, or "terrible lizards," a term first used in 1841. We, of course, now know that birds are dinosaurs, so in an ironic sense Hitchcock was correct; possibly we must give him credit for being right for the wrong reason.

FIGURE 3.11 Edward Hitchcock's "Paleontological Chart," showing geologically based, nonevolutionary trees for plants and animals, from *Elementary Geology* (1852).

Largely forgotten until quite recently (Archibald 2009), Hitchcock's trees of life seem quite odd to us for the simple reason that while showing the genealogy of various taxa, Hitchcock (and Agassiz) denied that evolution formed this genealogical pattern. Hitchcock maintained a staunchly antievolutionary stance throughout his life, even though he was clearly comfortable using a tree of life to show the history of life. The phrase "history of life" might seem incongruous for a creationist, but Hitchcock was no six-day, Young Earth creationist, for the simple reason that as a geologist he fully accepted an ancient Earth.

Figure 3.11 shows a black-and-white version of the hand-colored, 13- by 16-inch (33 by 41 cm) foldout chart from Hitchcock's eighth edition of *Elementary Geology* (1852). Other than slight color variations, this figure appears to be the same from the first edition (1840) until at least the thirtieth edition (1856). Unlike the chart, Hitchcock updated its explanation in later editions. Some salient excerpts, mostly from the first edition, are as follows, using Hitchcock's (1840) spelling, punctuation, and capitalization:

> In order to bring under the eye a sketch of the vertical range of the different tribes of animals and plants, that have appeared on the globe from the earliest times, the Chart which faces the title page, has been constructed. The whole surface is divided into seven strips, to represent Geological Periods. . . . The animals and plants are represented by two trees, having a basis or roots of primary rocks, and rising and expanding through the different periods, and showing the commencement, developement, ramification, and in some cases the extinction, of the most important tribes. The comparative abundance or paucity of the different families, is shown by the greater or less space occupied by them upon the chart. . . . The numerous short branches, exhibited along the sides of the different families, are meant to designate the species, which almost universally become all extinct at the conclusion of each period. . . . While this chart shows that all the great classes of animals and Plants existed from the earliest times, it will also show the gradual expansion and increase of the more perfect groups. The vertebral animals, for instance, commence with a few fishes; whose number increases upward . . . we ascend, until, in the Historic Period, the existing races, ten times more numerous, complete the series with MAN at their head, as the CROWN of the whole; or as the poet expresses it, "the diapason closes full in man." . . . In like manner if we look at that part of the Chart which shows the developement of the vegetable world, we shall see that in the lowest rocks, the flowering plants are very few. . . . Still more fully developed do we find them in the Historic Period; where 1,000 species of PALMS,—the CROWN of the vegetable world, have been found. (99–100)

Four of the geological "periods" in Hitchcock's diagram are no longer used: the primary corresponds to the Proterozoic aeon or Precambrian era; the Graywacke spans the Cambrian through the Silurian period; the Saliferous is the Triassic period; and the Oolitic, depending on the source, represents the middle and latter part or the entirety of the Jurassic period.

In describing his trees as "rising and expanding through the different periods, . . . showing the commencement, developement [*sic*], ramification, and in some cases the extinction, of the most important tribes [and t]he numerous short branches, exhibited along the sides of the different families, are meant to designate the species," Hitchcock seems to be describing an evolutionary process, but such is not the case. As noted, he certainly was neither a six-day, literal creationist nor a theistic evolutionist. Rather, Hitchcock regarded God's direct hand as the agent for biological change over long intervals of geologic time. Thus, unlike the English geologist Charles Lyell (before Darwin's influence), Hitchcock saw progression in the fossil record, which affected the way he represented it, providing us with a tree of life showing change toward the perfection of man—just below the angels (Archibald 2009).

Beginning with the thirty-first edition of *Elementary Geology*, which Hitchcock published with his son Charles from 1860 through 1866, the trees appear no more. The absence may be due simply to cost, but there may be a more directed explanation (Archibald 2009). From the first edition (1840) through at least the thirtieth edition (1856), Hitchcock attacked Lamarck's views on transmutation, as evolution then was often called:

> All the important classes of animals and plants are represented in the different formations. . . . Hence we learn that the hypothesis of Lamar[c]k is without foundation, which supposes there has been a transmutation of species from less to more perfect, since the beginning of organic life on the globe: that man, for instance, began his race as a monad, (a particle of matter endowed with vitality,) and was converted into several animals successively; the ourang outang being his last condition—before he became man. (91)

The next transmutation threat came in the form of Chambers's anonymously published *Vestiges of the Natural History of Creation* (1844). Hitchcock's first response to *Vestiges* was his inaugural address as president of Amherst College in 1845 (Lawrence 1972), but the first rebuttal in *Elementary Geology* came in the eighth edition (1847):

> [An] anonymous writer very strenuously maintains the doctrine of the creation and gradual development of animals by law, without any special creating agency on the part of the Deity. . . . But the facts in the case show us merely that the different animals and plants were introduced at the periods best adapted to their existence, and not that they were gradually developed from monads. In the whole records of geology, there is not a single fact to make such a metamorphosis probable; but on the other hand, a multitude of facts to show that the Deity introduced the different races just at the right time. (168)

In the coauthored edition (1862), there is no mention of *Vestiges*, and the trees of life no longer appear, but now Darwin assumes the mantle of transmutation bogeyman: "'We find in the history of fishes,' says Pictet, 'many arguments against the hypothesis of the transition of species from one into the other. . . . The connection of faunas, as Agassiz has said, is not material, but resides in the thought of the Creator.' It is well to take heed to the opinions of such masters in science, when so many, with Darwin at their head, are inclined to adopt the doctrine of gradual transmutation in species" (270).

The reference to so many being inclined to adopt the doctrine of transmutation suggests that Hitchcock saw his antievolutionary views waning. Further evidence comes from the fact that after thirty editions and twenty-two years, his *Elementary Geology* (1862) no longer carried his trees of life because in 1859 Darwin's *On the Origin of Species* appeared, and its sole figure—a foldout hypothetical tree of life—usurped this iconography to support evolution (Archibald 2009). In its place, we find a simple paleontological range chart of reptiles and amphibians modified from the anti-Darwinian Richard Owen's (1860) text on paleontology. A few thin lines connecting groups of reptiles from one geologic age to the next provide the only concession to the long history of life, and Hitchcock's exuberant trees of life have vanished.

A final pre-evolutionary tree offers a surprise: it represents the first tree—creationist or evolutionary—by a woman and may well be the only such tree by a woman until well into the twentieth century. Susan Butts (2010) relates the life and work of the author of this tree, Anna Maria Redfield (1800–1888, née Treadwell). Coming from a wealthy Canadian family, Redfield was well educated, including postgraduate classes in Clinton, New York, likely at what became Hamilton College. She was later honored with the equivalent of a master's degree from the now defunct Ingham University, the first institution of higher learning for women in the United States.

As with most wealthy Victorian women naturalists, Redfield amassed large collections of shells, minerals, botanical specimens, and scientific papers. While living in Syracuse, New York, she attended various scientific conferences and conventions, promoting *A General View of the Animal Kingdom* (1857), a wall chart showing in tree-like form the relationships of living animals, as well as her textbook *Zoölogical Science, or, Nature in Living Forms*, which first appeared in 1858 with at least five editions through 1874. Few Victorian women could claim the influence afforded by the publication and use of such an educational wall chart and textbook, yet Redfield remains a poorly recorded, minor figure in the history of women in science, and in the biological and evolutionary sciences (Butts 2010).

Redfield's remarkable wall chart, which measures just over 4 feet, 11 1/6 inches by 4 feet, 11 1/16 inches (1.5 by 1.5 m), shows considerable artistry in its representation of animal relationships in a clearly tree-like form (figure 3.12*A*). It remains unclear if she drew this complex lithographic diagram, although she mentions in the dedication to the first edition of *Zoölogical Science, or, Nature in Living Forms* (1858) that an "esteemed and highly competent friend" assisted. By 1865, this friend, the deceased Reverend E. D. Maltbie, is openly thanked. What role, if any, he

FIGURE 3.12 Anna Maria Redfield's (*A*) large, complex, tree-like wall chart *General View of the Animal Kingdom* (1857) and (*B*) similar but much smaller and simpler frontispiece of *Zoölogical Science, or, Nature in Living Forms* (1858).

A

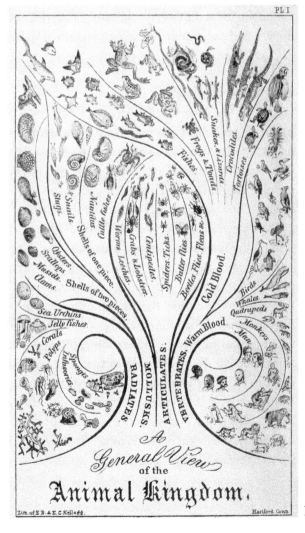

B

played in the production of the diagram remains unclear. In the diagram and in the much simpler version that serves as the frontispiece for all editions examined (see figure 3.12*B*), the organization follows Cuvier's four earlier nineteenth-century *embranchements*: Articulata, Radiata, Mollusca, and Vertebrata. These four main branches occur at the base of the wall chart; although unreadable at the scale provided, they clearly appear in the frontispiece diagram (compare figure 3.12*A* and *B*).

Redfield's first edition of *Zoölogical Science* was published a year before Darwin's *On the Origin of Species*, so the absence of a discussion of evolution as the basis for her tree might be understandable, but even in the editions of 1865 and 1867, no mention of the evolutionary ideas of Darwin or others appears. Recall that Hitchcock reacted strongly against successive iterations of evolutionary theorizing. Redfield appears consistent through at least the 1867 edition in her sparse but clearly antievolutionary statements. She makes the seemingly evolutionary assertion that in her tree, "each branch puts forth other branches bearing subdivisions" (10), yet later statements belie this sentiment. In her discussion of Owen's ideas on the muscles in the hands of apes and humans, she concludes, "The teeth, bones and muscles of the monkey decisively forbid the conclusion that he could by any ordinary natural process, ever be expanded into a Man" (22); and her views appear even stronger when writing, "There is no evidence whatever that one species has succeeded, or been the result of the transmutation of a former species" (482). Although unwittingly, the trees of people such as Agassiz, Hitchcock, and Redfield primed the public for accepting, if reluctantly, that trees of life might just represent evolutionary and not merely organizational visual metaphors of life.

One More Tree Before Darwin

In the latter half of the nineteenth century, the scientific community, if not the public, was ready for some explanation of why plants and animals appear to emerge from simpler forms. The evidence from geology, paleontology, biogeography, comparative anatomy, and embryology all pointed to the inevitability that evolution occurs. Dismissed by science but embraced by a curious public, the publication of the anonymous *Vestiges of the Natural History of Creation* (Chambers 1844) had primed the intellectual pump in England.

In France in 1850, the Academy of Sciences announced an essay prize competition for a study examining the laws governing how fossils are distributed through geologic time, the timing of appearances and disappearances of species, the relationships between the ranges of fossil species and geologic boundaries, and whether transformation or creation explained species' originations. Early submissions were not deemed worthy of the prize, so it was offered again, with a deadline in early 1856. The German paleontologist Heinrich Bronn (1800–1862) submitted an essay that was awarded the prize in 1861. It was published in German in 1858 and in French in 1861 with the German title translated as *Investigations into the Laws of Development of the Organic World During the Time of the Formation of Our Earth's Surface* (Gliboff 2007, 2008).

Both the German and French versions of Bronn's winning essay contained a hypothetical tree of life complete with letters explaining aspects of relationships shown on the tree, not unlike Darwin's (1960) famous tree in Notebook B and his much more elaborate hypothetical tree that was published in *On the Origin of Species* in 1859, the year after Bronn's tree appeared (figure 3.13*A*). Bronn's (1861) figure presents us with a wispy, tree-like figure labeled with letters. He appears to have been most concerned with addressing the idea that although there was a trend toward perfection, less perfect forms kept branching even after more perfected forms had appeared:

> Not only the animals without backbones, the Fishes, the Reptiles, the Birds with hot blood, the Mammals, and finally Man, appeared one after another, but . . . in lower kingdoms of the Radiates, the Mollusks, the Fishes, the highest branches of the system appeared only after the lower branches, however in such a way that the highest twig of a lower branch appears often later than the lower twig of a higher branch. One wants to represent this state of affairs by a figure, it is necessary to represent the system like a tree, where the more or less high position of the branches corresponds to the relative perfection of the organization, an absolute way and without holding account of the more or less high position of the twigs on the same branch. Thus . . . [the] first twig *g* of the fourth branch *D* appears only after it first twig *f* of the following branch *E* already appeared, etc. (899–900)

Although certainly not a creationist, Bronn was less accepting of Darwin's natural selection as the mechanism for species changes in particular, or any mechanism in general, to explain the pattern of change he saw in the history of life (Gliboff 2007, 2008; Williams and Ebach 2008).

Like Darwin, Bronn struggled to understand what caused the patterns of species distribution in time and space. Both Darwin and Bronn possessed knowledge in geology and paleontology; Bronn's was more accomplished in paleontology. Unlike Darwin, Bronn did not support the idea that species transformed over time and especially not by Darwin's natural selection, so the sudden appearances of species in the fossil record must result from some creative force. The cause was unknown to Bronn, although in his view new species always adapted to their ecologic surroundings while maintaining their general taxonomic groupings (a vertebrate remained a vertebrate, a mollusk remained a mollusk); but new forms explored new environments, so a sort of perfection and expansion of diversity occurred over geologic time. According to Bronn, more poorly adapted species decreased in numbers, leading to extinction, only to be replaced by the continuing creation of plants and animals, forming a succession of species over time. He argued that the successional fossil record documented this. In this way, species advanced or progressed over time, again as shown by the fossil record. Further, Bronn argued that the succession of fossils did not support the notions discussed earlier of embryologic development mirroring changes in major taxonomic groups (Gliboff 2007, 2008; Williams and Ebach 2008).

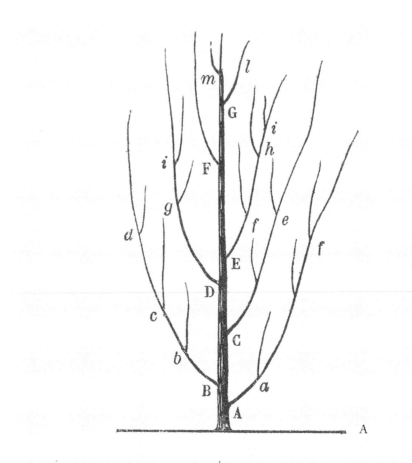

FIGURE 3.13 Heinrich Bronn's (A) hypothetical tree of relationships, from *Untersuchungen über die Entwicklungs-Gesetze der organischen Welt* (1858) and "Essai d'une résponse à la question de prix proposée en 1850" (1861), and (B) "Sequence of the Stratified Formations and Their Members and Distribution of the Organic Remains Therein," from *Lethaea geognostica* (1837–1838).

Darwin's book overshadowed Bronn's work, soon relegating his efforts to relative obscurity. Ironically, today Bronn (1860) often is best remembered for publishing the first German translation of Darwin's *On the Origin of Species*. Bronn's translation influenced the German biologist Ernst Haeckel, sometimes called the German Darwin, who reported having read Bronn's translation in German in the summer of 1860. As discussed in chapter 5, Haeckel was a consummate artist. Bronn, who experimented with visual methods of comprehending scientific data and ideas, also influenced Willi Hennig's visual representations of evolution. Bronn died in 1862, soon after the publication of his essay, and thus could contribute no more to the nascent field of evolutionary studies (Oppenheimer 1987; Richards 2004).

Interestingly, Hitchcock (1840) indicated in the footnote to his "Paleontological Chart" (see figure 3.11) that Bronn anticipated his tree figures in *Lethaea geognostica* (1837–1838). Based on Hitchcock's comment that Bronn produced "a chart constructed on essentially the same principles," one would expect to find a tree-like diagram, as in Hitchcock. This is most definitely not the case (see figure 3.13*B*). Translated from German, Bronn's figure is titled "Sequence of the Stratified Formations and Their Members and Distribution of the Organic Remains Therein." Like Hitchcock's chart, Bronn's figure shows deep geologic time, many lines representing many different groups known by fossils, and even some variation in line thickness to indicate relative taxonomic abundance. What Bronn's diagram does not show is any hint of a tree-like or branching diagram. His lines are unswervingly straight, with some change in thickness from bottom to top to indicate an increase in the number of species. What he produced is what paleontologists today refer to as a fossil range chart, which conveys when fossil taxa existed, not how they are related, except very generally by how they group on the diagram. That Hitchcock did not see much difference between his chart and Bronn's figure indicates that Hitchcock did not realize that his connecting the branches in a tree-like figure held any particular significance beyond what Bronn's unconnected lines showed (Archibald 2009). It is no wonder, then, that Hitchcock's tree-like "Paleontological Chart" disappeared from print once he realized that it was more akin to Darwin's (1859) hypothetical tree figure showing the evolution of life than to Bronn's diagram showing the paleontological succession of life.

CHAPTER FOUR

Deciphering Darwin's Trees

Charles Darwin did not discover evolution, a fact known to most modern biologists but not to many others. The idea of "descent with modification," or what we more colloquially call evolution, was a hotly debated topic by the early nineteenth century. The debate lacked what Darwin provided, along with Alfred Russel Wallace (1823–1913): a mechanism, and the mechanism was natural selection. After the publication of Darwin's *On the Origin of Species* in 1859, no one would ever look at a tree of life in the same way. Thus Darwin's various evolutionary trees, all but one of which went unpublished in his lifetime, deserve a closer look to provide a glimpse of what Darwin was pondering.

Darwin accomplished many things, but he did not draw well. This affected the way he chose to represent his ideas to the public and other scientists. It may explain why in all his publications, and they were numerous, he provided only one evolutionary tree, and that was hypothetical—the single figure in *On the Origin of Species*. Significantly, Darwin's trees deal more with his attempts to understand the process of evolution than its pattern. This is not surprising, inasmuch as he was convinced early on that evolution occurred, but he continued to struggle with the process of evolution and how best to represent it.

The other trees (and a few ancillary diagrams) discussed here, none published in Darwin's lifetime, date from his earliest musings in Notebook B through to his last known tree sketch, a phylogeny of primates that he drew in 1868. Discussion of the tree sketches follows a basically chronological schema, and, as will be seen, a clustering of his thoughts does affect in a general way the time intervals of the various groups of sketches. Exact times of drawing of some of the tree sketches can only be estimated to within a few years, whereas the others are known to the year, month, or even day of drawing.

Earliest Musings

Darwin's single most famous tree of life never appeared in print in his lifetime but became synonymous with Darwin because of its appearance in all manner of media leading up to, during, and following the year-long celebration in 2009 of the bicentennial of his birth and the sesquicentennial of the publication of *On the Origin of Species*. This branching stick figure appears on page 36 of his Notebook B, written in 1837 and 1838, but is not the first or only such figure in that volume. We must back up to page 26 in Notebook B to see the two earliest surviving diagrams by Darwin (1960) (figure 4.1). We must back up even further to see how Darwin reasoned before he drew trees on page 26. In the pages preceding these diagrams, he mused greatly on many aspects of evolution, but here I concentrate on those comments most specific to tree-like diagrams.

Starting in the middle of page 21 and continuing through page 26 of Notebook B, where the diagrams appear, Darwin (1960) writes:

> Changes not result of will of animal, but law of adaptation as much as acid and alkali. Organized beings represent a tree irregularly branched some branches far more branched—hence genera. —As many terminal buds dying as new ones generated [21]. There is nothing stranger in death of species than individuals If we suppose monad definite existence, as we may suppose in this case, their creation being dependent on definite laws, then those which have changed most owing to the accident of positions must in each state of existence have shortest [22] life. Hence shortness of life of Mammalia. Would there not be a triple branching in the tree of life owing to three elements air, land & water, & the endeavour of each ~~one~~ typical class to extend his domain into the other domains, and subdivision three more, double arrangement. —[23] if each main stem of the tree is adapted for these three elements, there will be certainly points of affinity in each branch A species as soon as once formed by separation or change in part of country repugnance to intermarriage ~~increases it~~ settles it [24] We need not think that fish & penguins really pass into each other. —The tree of life should perhaps be called the coral of life, base of branches dead; so that passages cannot be seen. —this again offers [25] contradiction to constant succession of germs in progress no only makes it excessively complicated. [sketch] Is it thus fish can be traced right down to simple organization. —birds —not? [sketch] [26]. (43–46)

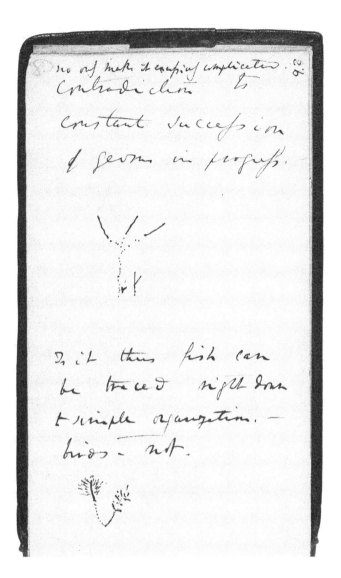

FIGURE 4.1 Two phylogenetic trees in Charles Darwin's Notebook B (1837–1838:26; Cambridge University Library MS.DAR.121.26). The top tree has a three-way split, showing Darwin's three environments of air, land, and water; the bottom tree bifurcates into a dotted line on the left (likely birds) and solid line on the right (likely fish). (Reproduced by kind permission of the Syndics of Cambridge University Library)

Paraphrasing Darwin's shorthand, adaptation rather than the will of a species results in changes that can be shown in an irregularly branching tree because branches differentially bud and die. Some species, such as monads, changed little on the tree, whereas others, such as mammals, changed greatly in part because of where by accident they occur on the tree. Life branched into three parts representing the conquest of the three great environments of air, land, and water. Each of these three main branches then invaded the other two environments. Certainly some of the species on the three branches resemble each other, such as fish and penguins, but because of a similar environment, not because of evolutionary relationship. Darwin continues that possibly a better metaphor for evolutionary relationships than a tree would be coral, in which most of the structure is dead and only the tips are alive. Such metaphors complicate our current understanding of how information is passed from one generation to the next. In such a diagram, fish can be easily traced to the bottom of the diagram, but birds cannot.

The first sketch on page 26 then follows (see figure 4.1, *top*). The meaning of the two smaller bifurcating figures is not clear, although possibly they imply seedlings or new corals. The larger, three-way splitting lines show Darwin's three environments of air, land, and water, in which the dotted lines indicate the dead parts and the solid lines the live parts of the branching coral of life. The sketch at the bottom of page 26 bifurcates into a dotted line on the left and a solid line on the right (see figure 4.1, *bottom*). A brush of small branches tops each line. Given what Darwin says immediately above the sketch, the left branch is likely the more poorly known evolutionary history of birds, whereas the right branch represents the better-known (especially fossil) fish.

The very next page in Darwin's (1960) notes supports this interpretation: "We may fancy according to shortness of life of species that in perfection, the bottom of branches deaden, so that in Mammalia, birds, it would only appear like circles, & insects amongst articulata. —but in lower classes perhaps a more linear arrangement" (44). By circles, Darwin certainly means what we term dotted lines.

On page 36 of Notebook B, we come to the now famous branching stick figure sometimes called the "I think" because these words appear at the top of the page (figure 4.2). Darwin (1960) writes above, to the right of, and below this diagram and continues on page 37:

> I think [*above*] Case must be that one generation then should be as many living as now. To do this & to have many species in same genus (as is) requires extinction [*right*]. Thus between A & B. immens gap of relation. C & B. the finest gradation, B & D rather greater distinction. Thus genera would be formed. —bearing relation [*bottom and top*, 37] to ancient types. —with several extinct forms for if each species an ancient (1) is capable of making 13 recent forms, twelve of the contemporarys must have left no offspring at all, so as to keep number of species constant. (45–46)

Darwin grapples with several ideas here. His diagram's circled 1 is clearly the ancestor in the sketch, the thirteen lines with crossbars are extant forms, and the twelve without crossbars are extinct forms. In today's usage of branching diagrams,

Darwin's *B*, *C*, and *D* form a polytomy indicating that they are equally related to one another, or, more correctly, that their relationships are not resolved. As Darwin correctly notes, *A* and *B* are more distantly related, but *A* is equidistant from *B* and *C*, but in Darwin's time and even until recently, the relative position on the tree and not just relative branching was intended to show closeness of relationship. He entertains the notion that over geologic time speciations should balance extinctions, something we now know may happen when environmental conditions remain somewhat stable, but at other times, especially times of mass extinctions, this balance skews to extinction and then often back to speciation. If anything, the total count of species has increased in the last 100 million years.

Interestingly, in the tree sketched on page 36, with the exception of one bifurcation (two splits), the remaining ten dividings form polytomies (multiple splits) of three or four branches. In all but a few of Darwin's trees shown in this chapter, polytomies of three or sometimes more branches far outnumber simple bifurcations. This may simply be habit, but based on the earlier quotes, it may be the case that Darwin viewed the process of speciation as forming more than two new lineages at each such event.

In some instances, Darwin does mention other researchers by name in Notebook B, but not in specific reference to the three diagrams on pages 26 and 36. Julia Voss (2010) suggests that when Darwin (1960) writes in

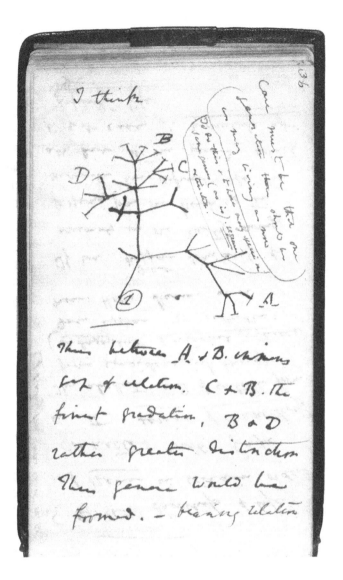

FIGURE 4.2 Darwin's well-known branching ("I think") hypothetical stick figure in Notebook B (1837–1838, 36; Cambridge University Library MS.DAR.121.36). (Reproduced by kind permission of the Syndics of Cambridge University Library)

Notebook B that "fish can be traced right down to simple organization," he may have had Louis Agassiz's tree-like paleontological chart for fossil fishes in mind (see figure 3.10). Parts of Agassiz's *Recherches sur les poissons fossils* appeared from 1833 through 1844. Voss gives the date for publication of the work as 1833, but the diagram did not appear until the final version in 1844 (Brown 1890; Archibald 2009). Thus Darwin could not have had this 1844 figure in mind while writing his 1837 notebook, although he could have been influenced by earlier work of Agassiz and others on the wealth of fish fossils compared with the essentially nonexistent fossil record for birds.

Voss (2010) and Theodore Pietsch (2012) speculate that the thirteen living species shown in figure 4.2 might represent the thirteen species of Galápagos finches that appear in the *Birds* volume of *The Zoology of the Voyage of H.M.S. Beagle* (1841). This somewhat lesser-known work was printed between 1838 and 1843 and often appears as a set of five bound volumes but originally was published as nineteen separate, unbound numbers dealing with fossil mammals, living mammals, birds, fish, and reptiles. It should not be confused with the commonly titled *Voyage of H.M.S. Beagle*, which was Darwin's first authored book and initially appeared in

1839 simply as *Journal and Remarks, 1832–1836*, the third volume of the four-volume *Narrative of the Surveying Voyages of His Majesty's Ships Adventure and Beagle, Between the Years 1826 and 1836, Describing Their Examination of the Southern Shores of South America and the Beagle's Circumnavigation of the Globe*. The captain of the *Beagle* and editor of these volumes, Robert FitzRoy, was none too pleased when Darwin's contribution became a great success with the public, appearing in many editions, whereas the other volumes languished. FitzRoy's vehement, ranting denunciation of Darwin's work on evolution overshadowed the memory of his later careers, checkered in diplomacy but distinguished in meteorology.

In contrast to his *Voyage of H.M.S. Beagle*, Darwin is usually credited with only editing and superintending the *Zoology* volumes, although he provided a preface and ecologic, geologic, and geographic information for some of the volumes. As anyone who has edited a book can testify, authors sometimes are a rather prickly and recalcitrant lot. Although John Gould receives credit as the author of the *Birds* volume of *Zoology*, Paul Barrett and Richard Freeman (Darwin 1987) note that Gould was in Australia collecting birds and mammals during most of the writing of the volume, much of which fell to Darwin with the assistance of others. Elizabeth Gould, John's wife, produced all fifty superb plates in the volume. Despite the problems with writing and production, Gould had informed Darwin about the importance of the Galápagos birds in general and the so-called finches in particular. I write "so-called" because in fact the newer evidence based largely on molecular analyses performed in part by my colleague Kevin Burns (Burns et al., 2002) supports the view that these birds are not finches but relate more closely to tanagers found in South America and the Caribbean. More to the point, we know that earlier in 1837, the year that Darwin sketched his now famous Notebook B tree, Gould had told him about the uniqueness of these Galápagos birds, but I am unaware of evidence pointing to thirteen species of finches being identified when Darwin began to write the earlier parts of Notebook B in July 1837.

The 1840s

As amply shown by John van Wyhe (2009), Darwin did not fear publishing his ideas on evolution from the late 1830s through the 1850s, as is often portrayed. Certainly, Darwin was still gathering information and developing his ideas on evolution, but his time during these years was consumed with preparing, editing, writing, and publishing eleven volumes (depending on how one counts) that deal with various aspects of the zoology and geology of the voyage of H.M.S. *Beagle* and fossil and recent barnacles, as well as over sixty scientific papers (Freeman 1977).

Tree-like diagrams in Darwin's notes from the 1850s cannot be pinned down as to year, but we do know the month and year of his two drawings from the 1840s. Both are small, obscure figures providing nothing profound, but they offer a glimpse of Darwin's struggles at the time regarding the process of evolution. The first figure, drawn in July 1843, appears almost as a doodled afterthought,

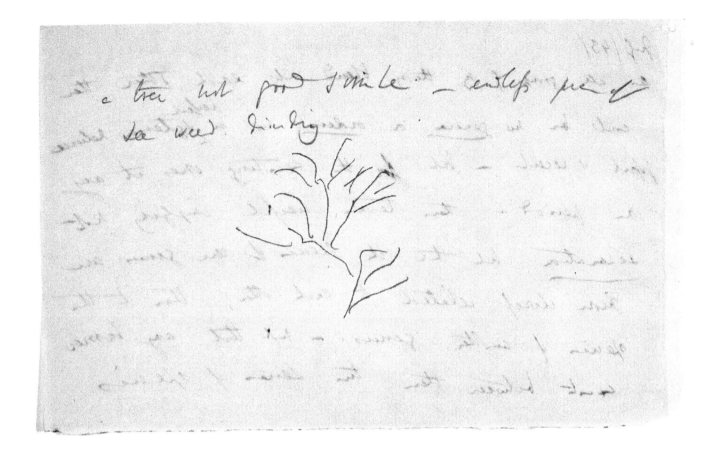

about six years after Darwin drew his tree-like figures in Notebook B. On the front side of a slip of gray paper, he writes:

> As all groups by my theory blend into each other, there could be no genera or orders «in same sense that no part or branch of a tree can be said to be distinct» in a «perfect» systema naturæ fossil & recent—but for the existing ones at any one period—these terms useful, implying not separation, but that the species of one genus are more closely related to each other, than to the species of another genus. —not that any barrier exists between these two series of species /over

On the reverse side, he writes, "a tree not good simile—endless piece of sea weed dividing" (Cambridge University Library MS.DAR.205.5.90r–v [angle quotes, or guillemets, are Darwin's later insertions]).

Darwin's seaweed sketch appears at the bottom of the reverse side (figure 4.3). Unlike any branching diagram he had done before or would do after, the wispy branches evoke the bending, curving form of a fragile, waterborne plant. One cannot know with certainty Darwin's thoughts here, but it would seem that he still struggled with the notion of all life through time as continuous and unbroken so that no metaphor—trees, corals, or seaweeds—captured his vision.

FIGURE 4.3 Darwin's wispy sketch from 1843 (Cambridge University Library MS.DAR.205.5.90v), with branches evoking the bending, curving form of a fragile, waterborne plant, suggesting the notion of life through time as continuous and unbroken seaweed. (Reproduced by kind permission of the Syndics of Cambridge University Library)

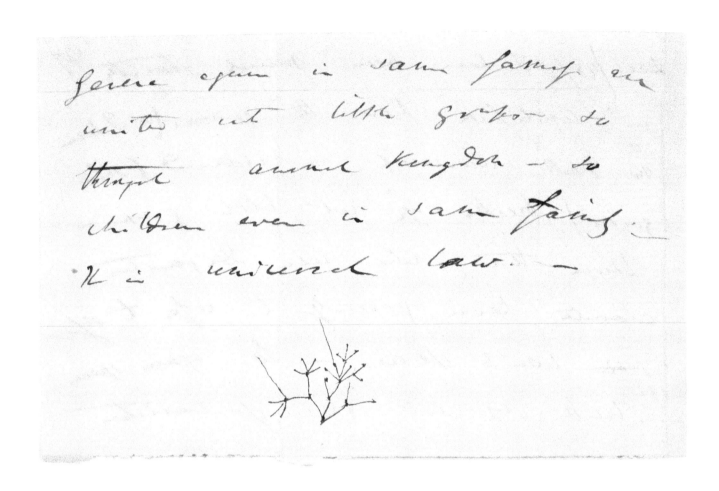

The second known drawing from the 1840s is on the reverse side of a small, off-white paper (figure 4.4). The front side reads:

Dec. /48/. I have been much struck in Anatifera how the genus, (& I have no doubt universal, as evidenced by sub-genera) breaks up into little groups—hence those who use Dignostic character have generally to refer to only 1 or 2 or 3 species—So again species break up into groups of varieties [*reverse side*] Genera again in same family are united into little groups—so throughout animal Kingdom—so children even in same Family.—It is universal law. (Cambridge University Library MS.DAR.205.5.127r–v)

Anatifera refers to what we now call *Lepas*, a genus of pedunculate barnacle that includes the species *Lepas anatifera* (Pedunculata, Cirrepedia), commonly called a goose or gooseneck barnacle. In letters to various colleagues, Darwin opined as to how he might handle the considerable taxonomic confusion surrounding the names *Anatifera* and *Lepas*. He had begun to work on his two monographs on barnacles: two volumes on fossil barnacles published in 1851 and 1854, and two volumes on extant barnacles published in 1851 and 1854. Both volumes from 1851 were titled *The Lepadidae, or, pedunculated cirripedes*, which includes *Lepas anatifera*.

The small branching diagram at the bottom of the page is reminiscent of the "I think" diagram on page 36 of Notebook B (see figure 4.2), except that instead of dashes terminating thirteen of the branches, Darwin here uses dots at the termini of some branches and at the nodes of others. Two single branches extend considerably beyond the others. No text or numbers appear on the tree-like sketch, so any suppositions as to what he meant are just that. The comments about *Anatifera* and other genera breaking into small groups suggest that he was trying to represent the idea of branches dividing into small groups, which in turn break into small groups in a polytomous fashion. The sketch in figure 4.4 suggests this, except that here the number of bifurcations (four) almost matches the number of polytomies (five). Given that Darwin began work on the barnacles in earnest in 1846 and the last of the four monographs appeared in 1854, eight years later, it is rather ironic that all we have preserved from this interval is this one small, rather enigmatic tree-like sketch. It does, however, show that Darwin continued to struggle with how species radiated and how this can be depicted.

The 1850s

Darwin's six tree sketches done in approximately the last eight years of the 1850s constitute the greatest concentration of his surviving tree sketches, including the only one published in his lifetime, the foldout figure in *On the Origin of Species*. The dates of production for four of the sketches cannot be established with any certainty, but Darwin probably drew one between 1852 and 1855, whereas the other three likely come from the year 1857 or 1858. Two very important firsts for Darwin appear in several of these sketches. On three of the four sketches, for the first time Darwin writes the names of actual animals, a feature seen in some of his later sketched trees. On the fourth sketch, he identifies geologic time on a tree for the first and only time. I find this somewhat unusual as until the publication of his barnacle monographs between 1851 and 1854, Darwin was known first and foremost as a geologist. Why, then, was geologic time not integral to his visual perceptions of how evolutionary relationships should be presented? Darwin did portray relative time and some ancestor–descendent relationships, but only once did he include geologic time.

The single sketch with geologic time intervals written on it shows five concentric circles, two of which are incomplete (figure 4.5). The inner to outer circles read "Palaeoz" for Paleozoic, "Second" for Secondary, and "Tertiary." As with Darwin's small tree sketch from 1848, polytomies—usually trifurcations—with only four bifurcations are most common. Also as with the 1848 sketch, small dots occur at most branchings and the tips of branches. They also appear at intervals along at least three branches. We know what they represent because at the top of the page Darwin writes "Dot means new form," but unfortunately the text that follows is not discernible with any certainty. Possibly the last word is "Bird." Darwin clearly had in mind what we today call cladogenesis for branching speciation and anagenesis when it appears that a species evolves into another without splitting.

It is quite likely that Darwin copied the idea of showing geologic time as a circle, possibly from Louis Agassiz and Augustus Gould's frontispiece for *Principles of Zoölogy* (1848) (Voss 2010) or from Heinrich Bronn's similarly shaped frontispiece for *Lethaea geognostica* (1850–1856) (see figure 7.18). Agassiz and Gould's

frontispiece, "Crust of the Earth as Related to Zoology," identifies a series of concentric, shaded circles as intervals of Earth history, with the oldest at the center. The circle divides radially into four quadrants identified as the four major animal *embranchements* first proposed by George Cuvier earlier in the nineteenth century: Articulata, Radiata, Mollusca, and Vertebrata. Recall that Agassiz studied with and was an acolyte of Cuvier. In this diagram, man rests at the top of the circle, embellished with a shining crown.

Each major group within each quadrant emanates as a ray from the center of the circle, showing its origin in the geologic past through to today, with trilobites and ammonites as the only extinct lineages. The relative thicknesses of the rays provide an estimate of their relative species richness at any given time. A dot at the center of the circle, according to Agassiz and Gould, represents "the primitive egg, with its germinative vesicle and germinative dot, indicative of the universal origin of all animals." Although clearly a nonevolutionary representation, the diagram invokes a comparison between the life history of an individual and the history of life on Earth, a theme strongly repeated in an evolutionary context some twenty years later by Ernst Haeckel (1866) in his coining of the terms "ontogeny" and "phylogeny." As Julia Voss (2010) argues, Darwin has pulled Agassiz and Gould's metaphor into an evolutionary realm by literally connecting the dots in his sketch (see figure 4.5). Very likely Darwin knew of this circular diagram, as he possessed a copy of Agassiz and Gould's book in his library (Cambridge University Library 1961).

Darwin uses the concentric circles for geologic time in one other tree sketch, dating most likely from 1857 or 1858, but without designating geologic times (figure 4.6). This represents one of seven trees by Darwin dealing with mammalian evolution (for six of them, see Archibald 2012). The top of the sketch reads, "Let dots represent Genera ???" To the right of the sketch, Darwin writes "no form intermediate." A line leads to the largest dot at the base of the tree or to the dotted lines below; thus "no form intermediate" almost certainly refers to the absence of species intermediate between marsupials and placentals (possibly rodents). To the left, the text probably reads, "If these had all given descendants then this wd [would] have been a great series," and a line leads to an encircled group of five dots near the base of the tree. He thus indicates that this cluster of dots did not radiate into many other species but would have been significant if they had done so. At the very base of the tree, Darwin writes, "Parent of Marsupials & Placentals," with the word "Rodents" written over the main middle trunk of the tree. Here are two groups that did very successfully radiate, as indicated by many connected dots. To the right, partial circles enclose the tree. Short lines labeled "Rodents" and "Marsupials" point to the middle and lower main trunks, respectively.

It at first may seem odd that Darwin should juxtapose rodents and marsupials in such a tree. Based on discussions and correspondence between Darwin and G. R. Waterhouse, there was some idea of a possible relationship between rodents and marsupials. This idea is reinforced in another, simpler diagram, also probably from 1857 or 1858, that appears on the reverse side of the paper bearing the tree shown in figure 4.5. The figure is labeled "Parents of Placentals Rodents & Marsupials" (figure 4.7). Waterhouse espoused the view that rodents in general,

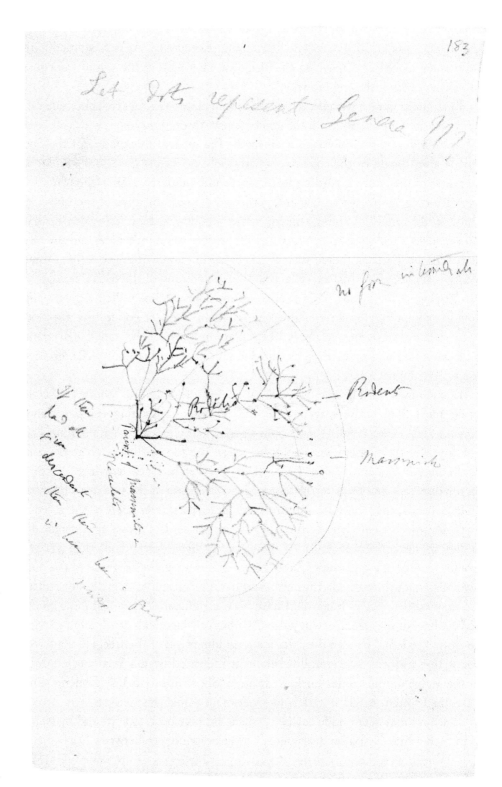

FIGURE 4.6 A phylogenetic tree of mammals, drawn by Darwin in 1857 or 1858 (Cambridge University Library MS.DAR.205.5.183r), showing placentals (specifically rodents) and marsupials as sister taxa. No other hypotheses of relationship are given. (Reproduced by kind permission of the Syndics of Cambridge University Library)

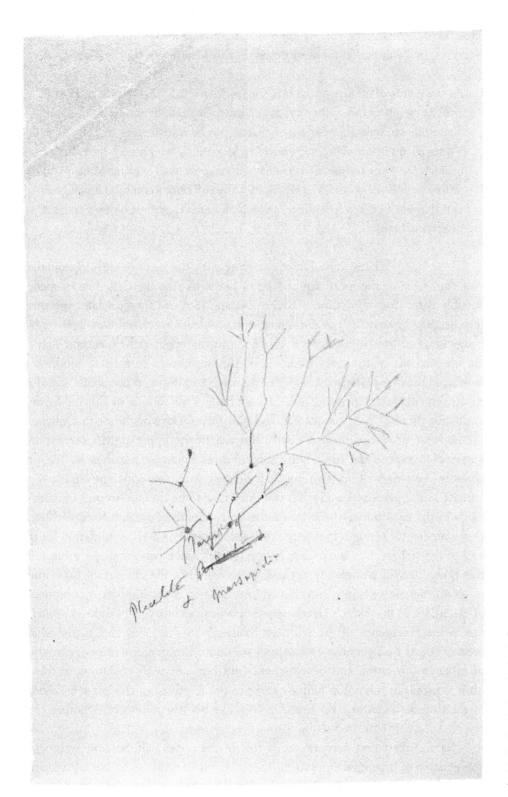

FIGURE 4.7 A simple phylogenetic tree, drawn by Darwin in 1857 or 1858 (Cambridge University Library MS.DAR.205.5.184v), indicating the "Parents of Placentals Rodents & Marsupials." (Reproduced by kind permission of the Syndics of Cambridge University Library)

and in particular the South American vizcacha (or bizcacha), a relative of the chinchilla, were somehow related to marsupials. This was largely due to the occurrence in both of a dual vaginal opening, which we know is an ancestral trait seen across various mammals and which evolved convergently in marsupials as a dual pseudovagina. Darwin (1859) even expresses this idea in *On the Origin of Species*:

> As the points of affinity of the bizcacha to Marsupials are believed to be real and not merely adaptive, they are due on my theory to inheritance in common. Therefore we must suppose either that all Rodents, including the bizcacha, branched off from some very ancient Marsupial, which will have had a character in some degree intermediate with respect to all existing Marsupials; or that both Rodents and Marsupials branched off from a common progenitor, and that both groups have since undergone much modification in divergent directions. (430)

These tree sketches show Darwin's toying with the ideas of early mammalian evolution. One of the more astounding of Darwin's tree sketches done in pencil possibly dates from sometime between 1852 and 1855 and thus predates the three sketches just discussed. It is not a single tree but five trees suffused later with a hodgepodge of brown ink notes overlapping in the upper-right quadrant that all but obscure one of the trees (figure 4.8). The impression of mental doodling is reinforced because Darwin drew this busy sketch and notes on the blank side of an advertisement for the "printing office and stationery warehouse" of Edward Strong, located in Bromley, Kent, about 5 miles (8 km) from Darwin's home in Downe.

For ease of discussion and clarity, the text in the upper-right quadrant was removed to expose the underlying tree, and, as much as possible, legible text replaces Darwin's writing (figure 4.9). Emerging from this apparent jumble is an exercise in the perceived, and what we now know to be the importance of embryology in the evolutionary process. At the time, scientists, notably Richard Owen, were pressing for changes in embryological development as the foundation for the idea of the archetype as the underpinning for differences among major groups. By this time, Darwin was clearly becoming convinced of the efficacy of his natural selection, but he certainly entertained embryology as an important component of evolution, as this sketch clearly shows. Ironically, embryology did not contribute to the emergence of the Modern Synthesis in the 1930s and 1940s, which melded population genetics and natural selection. Embryology in its incarnation of evodevo (evolution and development) did not join with evolutionary theory until the tools of molecular biology came to the forefront in the last few decades of the twentieth century. Yet here Darwin, like his late-nineteenth-century contemporaries, contemplated its importance.

Darwin likely first drew the large tree in the upper half, possibly writing an abbreviation of "mammal embryo" at its base. He then scratched it out with a series of pencil swirls, to be replaced with the figure just below, showing a tree of mammalian carnivores on the left and what loosely can be called mammalian herbivores on the right. The dual placement of "dog" on the tree was certainly no mistake, as he favored the idea that domestic dogs likely descended from a series of ancestors

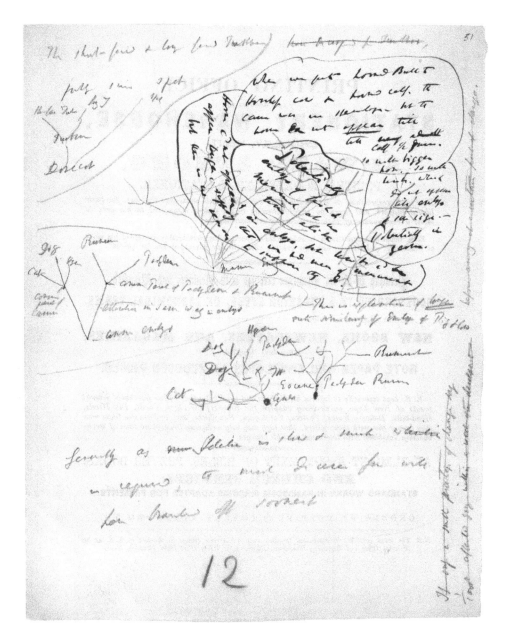

FIGURE 4.8 Five tree sketches, drawn by Darwin sometime between 1852 and 1855 (Cambridge University Library MS.DAR.205.6.51r), with overlapping notes in the upper-right quadrant all but obscuring one of the trees. (Reproduced by kind permission of the Syndics of Cambridge University Library)

within the canid family. This tree was probably followed by the simpler version on the left that labels various embryos in ancestral positions. At some point, he added his then emerging information of pigeon breeding at the top left in a tree sketch. Embryos are not mentioned in this tree, but likely he related artificial selection in pigeons to changes in embryological development in the different pigeon breeds. The crossed-out comment on the far right also speaks to early stages of embryological development as an engine for evolutionary change.

Next we must deal with what Darwin penned in brown ink over the scratched-out intricate tree in the upper half (see figure 4.8 [transcriptions mostly from Charles Darwin Papers]). The uppermost and lower-right bubbles read, "When we put horned Bull to Hornless cow & horned calf. the cause was in embryo [?]—but the horns do not appear till ~~nearly adult~~ calf ½ grown. so with bigger horn,

The short-faced & long faced Tumbler   have diverged from Tumbler,

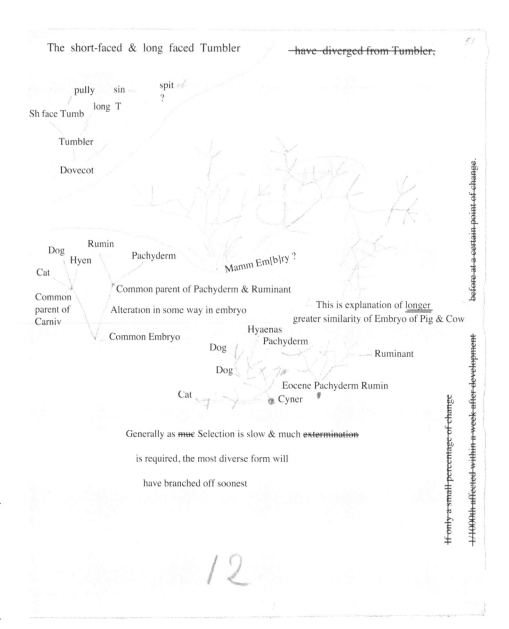

FIGURE 4.9 The same five trees shown in figure 4.8, with the over-written text removed and transcriptions of Darwin's written comments added. (The transcriptions are mostly from Charles Darwin Papers.)

so with link, which do not appear till embryo of same size. —Potentially in germ." The middle bubble reads, "Potentially embryo of fish & Mammal at no [or one?] time[s?] alike." The lower-left bubble reads, "Horns c^d not appear & in embryo, but limbs c^d be appear longer, supposing that we had means of measurement but there is no reason to suppose they do."

These penned comments overlying the tree sketches demonstrate Darwin's struggle with understanding how traits are passed from one generation to the next. It is tempting to suggest that Darwin comes close to understanding the underly-ing germ-line inheritance of traits. This was not to be, for in *The Variation of Ani-mals and Plants Under Domestication* (1868), he published his theory of "pangenesis," which argued that "gemmules" passed from the soma, or body, of the organism to the gonads and thus to the next generation. We now know this not to be the case, as it necessitates the ideas of "use and disuse" and "inheritance of acquired characteristics," which Lamarckian evolutionary theory included. His ideas of inheritance never caught hold, but he never abandoned them, even mockingly calling them "my much despised child" in a letter to E. Ray Lankester in 1870 (Darwin 2010:69).

Although any specifics from the stream-of-consciousness jottings on the tree sketches in figure 4.8 may not be very important, Darwin does provide his general impressions about the importance of embryology to evolutionary change and how traits pass from one generation to the next. The tree sketches merely served as a vehicle for these musings.

Darwin's "Big Book"

By 1858, Darwin had completed some ten and a half chapters for his planned large work, or as he sometimes called it, his "Big Book" dealing with evolution by means of natural selection. Except for the first two chapters, which grew into *The Variation of Animals and Plants Under Domestication* (1868), the remainder of the manuscript was not published until 1975, when it appeared as an edited volume, *Charles Darwin's Natural Selection: Being the Second Part of His Big Species Book Written from 1856 to 1858*. This work includes two diagrams dealing with biological relationships: Darwin called one a table, and the other forms a multipage precursor to the tree that appears in *On the Origin of Species*. This precursor tree (actually trees) did not see wider circulation until its publication in *Natural Selection* (Darwin 1975) and still remains rather obscure.

Darwin's table in the same work, as well as two other related figures showing biological relationships and hybridizations, warrant brief comment. These three sketchy diagrams attempt to show the importance for evolution that Darwin placed on what he most often referred to as "hybridism." Darwin felt strongly that hybridization must play a role in evolution, such as in the formation of new species, but not until the advent of more molecularly based tools in the mid-twentieth century could this be demonstrated (see chapter 7). Darwin (1975) describes his table as providing for "all the well authenticated crosses which I have heard of in one order of Birds, the Rasores [no longer recognized]; in order that those who have not attended to the subject, may see how numerous the crosses have been, & between what different forms" (434). Figure 4.10*A* presents Darwin's original sketch (a more readable version is in Darwin 1975:434). Various birds referred to as the then recognized Rasores, such as pheasants and peacocks, are listed down the middle of the table, with brackets on either side linking species known to hybridize. Darwin indicates, "The Brackets, imply that hybrid offspring has been produced by the two forms so connected" (436). Small numbers barely discernible next to the species names refer to several pages of footnotes documenting the degree of fertility in the hybrids. This diagrammatic table is not tree-like; two other of Darwin's unpublished diagrams explore hybridization in birds, one of which at first appears tree-like, whereas the other uses some form of unrooted networks.

Stylistically, unrooted networks still exist, mostly for the comparison of populations within a species, usually by means of molecular techniques. Unrooted networks for systematic research reached their heyday in the 1960s and 1970s under the auspices of phenetics (see chapter 7). Unrooted networks have an even longer run, both before and after Darwin. For example, the English naturalist Hugh Edwin Strickland (1811–1853) published what he termed a map on which connecting lines of various lengths indicate relationships of birds (Strickland 1841; Voss 2010). In 1856, Alfred Russel Wallace used Strickland's technique to create an unrooted network to show the relationships within the now rejected avian order Scansores (birds with two toes facing forward and two facing backward). While it is tempting to claim that Strickland's and Wallace's unrooted networks represent pre-Darwinian evolutionary trees, nothing of their form or in the text indicates evolution as the basis for the arrangements. Rather, these visual schemes cleverly present a rather older heritage of unrooted networks useful in representing relationships.

Of the two additional Darwinian diagrams noted earlier, the first one appears to be some sort of tree and the other appears to be an unrooted network, but both more likely represent Darwin's attempts at creating a bracketing scheme in order to join hybridizing pairs of birds. Figure 4.10B, done sometime between the end of September and the end of December 1857, shows crosses between various species of mostly birds in the pheasant family listed down the middle of the figure. What appear to be branching figures or trees on either side more likely show nested brackets pointing to hybridizing pairs. Darwin drew lines through the text on this page as well as through the figure, indicating his rejection of this attempt. A probably later effort, shown in figure 4.10C, reads across the figure in Darwin's hand, "Make Table of Pheasants & Fowls Crossing," likely a note to himself. It presents Darwin's experimentation possibly with an unrooted network diagram. Although the intent of his artistry may not be clear, its content attempts to show some aspects of hybridization. If Darwin's comment about a table refers to the table that eventually appeared in *Natural Selection* in 1975, Darwin possibly had abandoned an unrooted network for a simpler bracketing version.

The relationship of each of these three hybridization diagrams to tree-like figures must remain tangential, but not so Darwin's large, carefully prepared foldout tree (actually trees) finally published in *Natural Selection* in 1975. Its publication was, of course, beaten by well over one hundred years with the appearance of the single, foldout figure in *On the Origin of Species*, published in 1859. The two sets of diagrams are similar, with the *Natural Selection* version more complex. The fact that it opens downward rather than upward, as in *On the Origin of Species*, seems inconsequential. Darwin identified the *Natural Selection* figures as comprising

FIGURE 4.10 Three diagrams drawn by Darwin showing his views and evidence for hybridization in various groups of birds: (A) a self-described table from 1858 indicating well-authenticated crosses for birds within the then recognized order Rasores (Cambridge University Library MS.DAR.12.88); (B) an eventually rejected figure (note diagonal pencil lines) from 1857 showing crosses between various species of birds mostly in the pheasant family (MS.DAR.205.7[1].33v); and (C) a figure likely from 1857 or 1858 showing various crosses of birds over which Darwin has written "Make Table of Pheasants & Fowls Crossing," possibly referring to the table in (A) (MS.DAR.205.7[1].86r). (Reproduced by kind permission of the Syndics of Cambridge University Library)

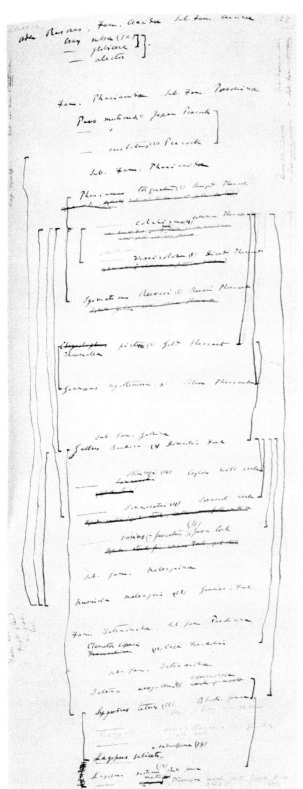

A

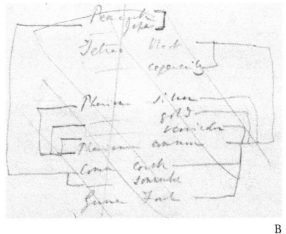

B

C

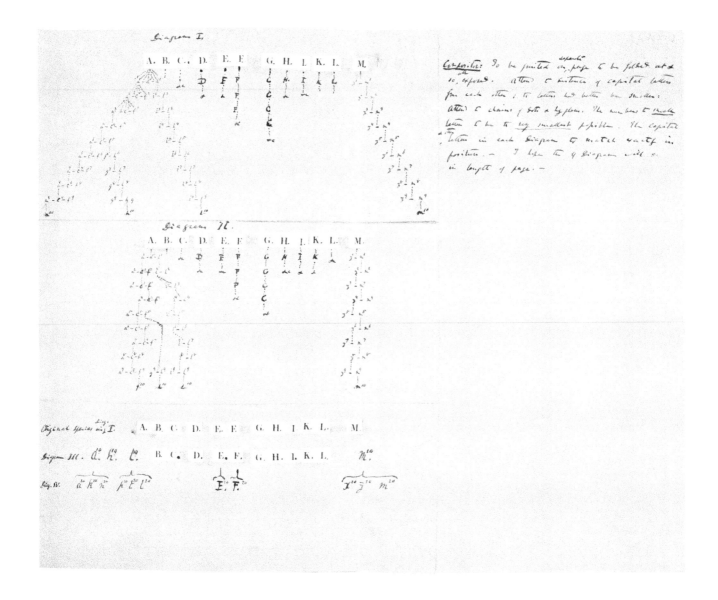

four diagrams, labeled I through IV (figure 4.11). The text in the upper-right quadrant details how the compositor was to arrange lettering on the diagram. Darwin expended some 3,500 words explaining these four diagrams, whereas for the single diagram in *On the Origin of Species* (figure 4.12), he used some 2,700 words in explanation. These texts allow us to understand more clearly what Darwin attempted to tell us about his ideas compared with earlier tree sketches. None of the trees in these two books tells us the specific patterns of relationship among plants and animals, but they attempt to show how Darwin perceived the process of evolution to unfold. About the diagrams in *Natural Selection*, Darwin (1975) writes, "The complex action of these several principles, namely, natural selection, divergence & extinction, may be best, yet very imperfectly, illustrated by the following Diagram, printed on a folded sheet for convenience of reference. This diagram will show the manner, in which I believe species descend from each other & therefore shall be explained in detail: it will, also, clearly show several points of doubt & difficulty" (238–39).

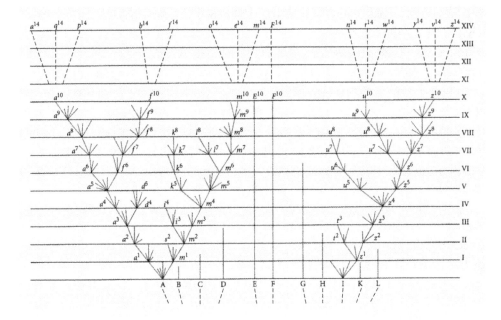

FIGURE 4.12 Darwin's only phylogeny published in his lifetime
appears in *On the Origin of Species*
(1859) as a foldout inserted
between pages 116 and 117.

Darwin's explanation, although rather long-winded, summarizes his several points. In diagram I, *A* through *M* are species of plants within a single genus with the species *A* and *M*: "the two most distinct forms in all respects. . . . From our principle of divergence, the extreme varieties of any of the species, & more especially of those species which are now extreme in some characters, will have the best chance, after a vast lapse of time, of surviving; for they will tend to occupy new places in the economy of our imaginary country" (239). Based on modern studies, the argument that extreme forms tend to survive is highly questionable, but Darwin does show how species may diverge through time.

For the sake of argument, Darwin (1975) makes "A the most moisture loving & M the least moisture-loving species." In his thought experiment, he then evolves each of the two species in what he sees as the most likely scenarios:

> We will first take the simplest case. Let M inhabit a continuous area, not separated by barriers, & let it be a very common & widely diffused & varying plant. From the fact of M. being very common & widely diffused, it clearly has some advantages in comparison with most of the other inhabitants of the same country . . . it could endure still more drought. . . . As m^1 tends to inherit all the advantages of its parent M, with the additional advantage of enduring somewhat more drought, it will have an advantage over it, & will probably first be a thriving local variety, which will spread & become extremely common & ultimately, supplant its own parent. We may now repeat the process . . . perhaps many thousands of generations may pass before m^1 will produce another variety m^2, still more drought-enduring & yet inheriting the common advantages of m^1 & M. . . .
>
> In each stage of descent, there will be a tendency in the new forms to supplant its parent, though probably, as we shall see, very slowly, & so ultimately cause its extinction. (239)

The successive replacement within the M lineage in Darwin's diagram refers to what is now called anagenesis. The process that Darwin envisions for species A is now called cladogenesis for obvious reasons—it results in the increase of the number of clades and species. Here Darwin shows mostly cladogenesis with some anagenesis but with the result of three species at the end where there had been one at the beginning. Unlike for M, Darwin notes that A varies greatly, giving rise to many varieties and eventually species. The printed capital letters and ink sketching appear clearly in figure 4.11. Less discernible faint solid and dotted pencil lines radiate from many of the species emanating from A, likely added by Darwin to emphasize the splitting nature of speciation in A as compared with M.

The other ten species, B through L, which at first glance appear to have become extinct, in fact survive, but Darwin has placed at each terminus only "&" or "&c," indicating the continuation of each of these lines, for he notes, "The other species of the genus, B to L, are supposed to have transmitted unaltered descendants" (244). This essentially defines what in the twentieth century Niles Eldredge and Stephen Jay Gould (1972) included as the equilibrium part of their theory of punctuated equilibrium.

The only difference that Darwin (1975) wishes to impart between diagram I and diagram II is that in diagram II "everything is the same as in diagram I . . . except that it is left to mere chance in each stage of descent, whether the more or less moisture loving varieties are preserved" (244), with the result that he shows the descendants in diagram II as less separated from each other on the horizontal scale compared with those in diagram I. He has argued, then, that natural selection will cause more divergence than chance alone, an idea that finds support in modern population studies. Notably, Darwin did not pencil in any additional radiations of species emanating from A in diagram II, as he had done in diagram I.

Although diagrams III and IV are not trees, comments are germane. As figure 4.11 shows, M is farther away from L than L is from K in all four diagrams. These differences have been lost in printed reproductions of this figure, such as in *Natural Selection* (Darwin 1975), that show equal spacing; thus Darwin's intent is lost. Darwin states, "So again m^{10} having constantly diverged from the characters of M will now stand more distant from L, than M originally stood. This is represented in the Diagram III" (245). He says that the descendants will go on to produce "more & more new specific forms & thus more & more modified or divergent." In diagram IV, Darwin attempts to show, not very successfully, the principle of what he terms analogy and what we now call convergence. He further notes that if all the intermediates became extinct, the remaining forms in diagram IV might be placed in separate genera.

A final interesting point Darwin (1975) makes about his diagrams in *Natural Selection* deals with the irksome question of the origin of higher taxa, such as families, orders, and classes:

> Now for a moment let us go back many stages in descent: on our theory the original twelve species A to M are supposed to have descended & diverged from some one species, which may be called Z, of a former genus. . . . Z will have become the ancestor of two or three very distinct groups of new species; & such groups, naturalists call genera. By continuing the same process,

namely the natural selection of generally the most divergent forms, with the extinction of those which have been less modified & are intermediate, Z may become the ancestor of two very distinct groups of genera; & such groups of genera, naturalists call Families or even Orders. (246)

Rush to Publish

Understandably, the publication of *On the Origin of Species* in 1859 heralded the acceptance of trees of life into the scientific as well as the public psyche. Never mind that the sole figure in Darwin's work was not a tree of life that showed the relationships of any specific species to one another but a tree that hypothesized Darwin's views of how evolution may have unfolded. The book provided the archetypical watershed moment when people acceptingly visualized nature as a great tree with its numerous branchings and rebranchings. A tree now was *the* metaphor for the history of life.

Even though Darwin does not mention human evolution in *On the Origin of Species*, the publication engendered a truly world-changing perception of nature and our place in it, for which reason one might conclude that his foldout tree became an icon of biology. But such was not the case. Indeed, today the tree is not even universally recognizable among biologists. Darwin's (1960) small figure from Notebook B is far more recognized and reproduced, especially since the celebration of the bicentenary of his birth in 2009 (see figure 4.2). This tree, not the one in *On the Origin of Species*, has come to symbolize Darwin's evolutionary views.

The foldout diagram in *On the Origin of Species* (see figure 4.12) both resembles and differs from those in *Natural Selection* (1975). In *Natural Selection*, as in *On the Origin of Species*, the "species are supposed to resemble each other in unequal degrees, as is so generally the case in nature, and as is represented in the diagram by the letters standing at unequal distances" (Darwin 1859:116). In the latter, Darwin has added fifteen horizontal lines with the explanation, "The intervals between the horizontal lines in the diagram, may represent each a thousand generations; but it would have been better if each had represented ten thousand generations. After a thousand generations, species (A) is supposed to have produced two fairly well-marked varieties, namely a^1 and m^1. These two varieties will generally continue to be exposed to the same conditions which made their parents variable" (117). He continues discussion a few pages later:

> If we suppose the amount of change between each horizontal line in our diagram to be excessively small, these three forms may still be only well-marked varieties; or they may have arrived at the doubtful category of sub-species; but we have only to suppose the steps in the process of modification to be more numerous or greater in amount, to convert these three forms into well-defined species: thus the diagram illustrates the steps by which the small differences distinguishing varieties are increased into the larger differences distinguishing species. (120)

For Darwin, then, the same processes—notably, natural selection—act over all time frames and eventually over all taxonomic levels to produce varieties and species to orders and classes. Only the length of time provides the variable.

In *On the Origin of Species*, Darwin does not introduce a hypothetical case, such as the moisture-loving through drought-resistant plants he introduced in *Natural Selection*, but instead remains abstract in his discussion. He still argues that "the more diversified in structure the descendants from any one species can be rendered, the more places they will be enabled to seize on, and the more their modified progeny will be increased," but here he also now admits, "I am far from thinking that the most divergent varieties will invariably prevail and multiply: a medium form may often long endure, and may or may not produce more than one modified descendant" (119).

The dashed lines at the top of the foldout in *On the Origin of Species* represent a more abstracted continuation of divergence and speciations shown lower in the diagram. In this tree, species *I* diverges but to a lesser degree than species *A*, but Darwin does not offer any statements as to differences or significances for species *I*. He simply calls it a second species following "analogous steps." After referring to species *I*, he does, as in *Natural Selection*, speak to the lack of divergence in some species: "The other nine species (marked by capital letters) of our original genus, may for a long period continue transmitting unaltered descendants; and this is shown in the diagram by the dotted lines not prolonged far upwards from want of space" (121). Following this on the same page, he notes that extinction is an important principle: "Hence all the intermediate forms between the earlier and later states, that is between the less and more improved state of a species, as well as the original parent-species itself, will generally tend to become extinct." The remainder of Darwin's discussion of the foldout covers in detail a point made in his discussion in *Natural Selection*—that the divergence of lower taxa such as species eventually leads to higher taxa such as orders and classes.

Darwin (1859) concludes the chapter on natural selection and his description of his foldout diagram by paying homage to the metaphor of a tree representing the history of life. It bears repeating in its entirety:

> The affinities of all the beings of the same class have sometimes been represented by a great tree. I believe this simile largely speaks the truth. The green and budding twigs may represent existing species; and those produced during each former year may represent the long succession of extinct species. At each period of growth all the growing twigs have tried to branch out on all sides, and to overtop and kill the surrounding twigs and branches, in the same manner as species and groups of species have tried to overmaster other species in the great battle for life. The limbs divided into great branches, and these into lesser and lesser branches, were themselves once, when the tree was small, budding twigs; and this connexion of the former and present buds by ramifying branches may well represent the classification of all extinct and living species in groups subordinate to groups. Of the many twigs which flourished when the tree was a mere bush, only two or three, now grown into great branches, yet survive and bear all the other branches; so with the species

which lived during long-past geological periods, very few now have living and modified descendants. From the first growth of the tree, many a limb and branch has decayed and dropped off; and these lost branches of various sizes may represent those whole orders, families, and genera which have now no living representatives, and which are known to us only from having been found in a fossil state. As we here and there see a thin straggling branch springing from a fork low down in a tree, and which by some chance has been favoured and is still alive on its summit, so we occasionally see an animal like the Ornithorhynchus or Lepidosiren, which in some small degree connects by its affinities two large branches of life, and which has apparently been saved from fatal competition by having inhabited a protected station. As buds give rise by growth to fresh buds, and these, if vigorous, branch out and overtop on all sides many a feebler branch, so by generation I believe it has been with the great Tree of Life, which fills with its dead and broken branches the crust of the earth, and covers the surface with its ever branching and beautiful ramifications. (129–30)

Marsupials: Sister Taxa or Ancestors?

Perhaps the most intriguing of Darwin's trees, or actually pair of trees, comes from a ten-page letter that Darwin wrote to the geologist Charles Lyell on September 23, 1860, during Darwin's vacation with his family at a seaside hotel in Eastbourne. The letter was written just ten months after the publication of *On the Origin of Species*. Its interest lies first in the fact that unlike all but a few of Darwin's trees, it details the relationships of an actual group—mammals. Second, it is a pair of trees with competing views on the origin of placental and marsupial mammals, something Darwin never did before or after, as far as we know. Third, unlike most of his other trees, except those in *On the Origin of Species* and *Natural Selection*, it comes with a commentary of his views, sometimes in rather frank terms, because it was meant for only Lyell's eyes (for details of Darwin's exchange of letters with Lyell, see Archibald 2012).

The two trees in question lay out two schemes for the origin of the two great clades of living therian mammals: marsupials and placentals (figures 4.13 and 4.14). Here the concentration concerns the two phylogenies and what Darwin said about them. At the top of the first tree, labeled "Diagram I," Darwin writes, "A. Unknown form probably intermediate between reptiles mammals, Reptiles & Birds as intermediate as Lepidosiren now is between Fish and Beatractians [Batrachians],—probably more This unknown form probably more closely related to Ornithorhynchus than to any other known form" (see figure 4.13). On the opposite side, labeled "Diagram II," Darwin writes "A (as on other side)." On both diagrams, an *A* was positioned at the origin of each tree; thus the description was intended to be the ancestor in both trees.

The two main branches of diagram I (see figure 4.13) are more symmetrical than those in diagram II (see figure 4.14), and the relative lengths of the branches

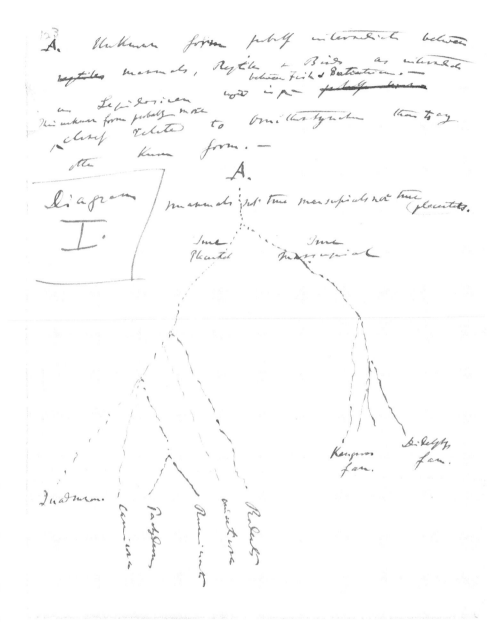

FIGURE 4.13 A phylogenetic tree, drawn by Darwin in 1860, showing a nonmarsupial, nonplacental mammal as the common ancestor, A, of both marsupials and placentals. (Reproduced with permission of the American Philosophical Society)

vary slightly. The topology or relative positions of the branchings in the two diagrams appears almost identical, except that the first node in the marsupial side is reversed. The differences appear inconsequential in the positioning and orientation of terminal taxa names. Names listed in each tree at the tips of terminal branches are identical, except for minor differences such as capitalizations.

In diagram I, on the "True Marsupial" branch, "Kangaroo fam." and "Didelphys fam." refer to what we today essentially recognize as the families Macropodidae and Didelphidae, respectively, although the content of each group in Darwin's time was somewhat broader. On the "True Placental" side, the first branch is "Rodents" with "Insectivores" as the second branch. The orders Rodentia and Carnivora are recognizable. Insectivora tended to include any small placental mammals with sharp teeth that could not be aligned with any other placentals. Consequently, Insectivora

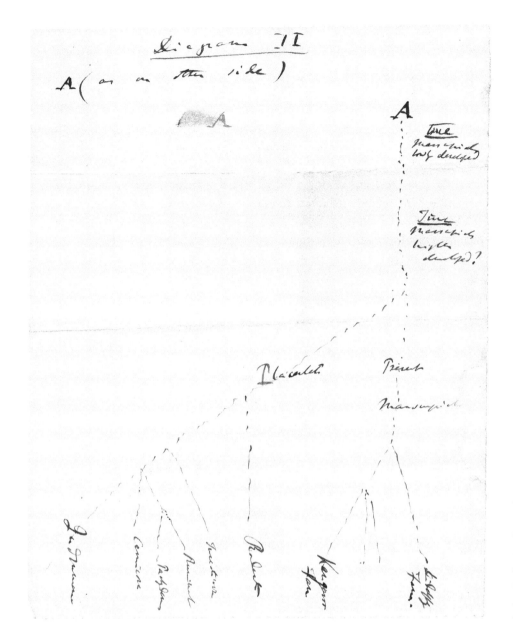

FIGURE 4.14 A phylogenetic tree, drawn by Darwin in 1860, showing a marsupial as the common ancestor, A, of both marsupials and placentals. (Reproduced with permission of the American Philosophical Society)

constitutes what biologists term a wastebasket taxon, and current classification utilizes the name Lipotyphla or Eulipotyphla, including shrews, moles, hedgehogs, and a few lesser-known groups. There is then a three-way split between one branch with "Ruminants" and "Pachyderms" (thick-skinned mammals), the second for "Carnivores," and the third for "Quadrumana." Johann Friedrich Blumenbach (1779) proposed "Quadrumana" (referring to four hands) for all primates except humans in contradistinction to his "Bimana" (two hands) for humans. Darwin here simply used this older terminology. Thomas Henry Huxley (1863) argued convincingly that the so-called higher apes were reasonably put in Carl Linnaeus's (1758) Primates along with humans. The names Quadrumana and Bimana were out of favor by the end of the nineteenth century in favor of Primates, including humans.

The most interesting aspects of the two trees are not Darwin's views of the relationships within marsupials and within placentals but the alternative hypotheses he presented regarding the origin of marsupials and placentals (Archibald 2012). On diagram I above the split between "True Placental" and "True Marsupial," Darwin writes "Mammals not true marsupials not true placentals" (see figure 4.13). On diagram II, the earliest ancestor is labeled "true Marsupials lowly developed" followed by "True Marsupials higher developed?" with the split leading to "Placentals" and "Present Marsupials" (see figure 4.14).

In the letter to Lyell accompanying these diagrams, Darwin indicated that as a general rule he preferred the tree in diagram I, but if the embryological brain of marsupials closely resembled that of placentals, he should strongly prefer the tree in diagram II, as this agreed with the antiquity of *Microlestes*. We now know that the marsupial embryologic brain does not indicate its being a precursor to the placental brain and that the Cretaceous fossil *Microlestes* (now *Thomasisa*) is not a marsupial.

Although it would be tempting to suggest that Darwin favored one of the trees over the other, especially if it were the correct version shown in diagram I, the evidence in his letter and diagrams is not definitive on this matter. The best we can argue from current evidence is that Darwin did weigh alternative views for the monophyletic versus the successive origin of living mammals, which no one else was doing with the same acumen 150 years ago (for further discussion, see Archibald 2012).

Man's Place in Nature

Darwin drew two phylogenetic trees dealing with primate evolution in general and human evolution in particular, neither published in his lifetime. The simpler and probably earlier of the two trees appears as a pencil sketch on the reverse of an undated, nicely handwritten ink text with scrawled pencil comments at the bottom (Cambridge University Library MS.DAR.80.B118r). The portion of the text in ink almost certainly belongs to Charles Darwin's daughter Henrietta Emma "Etty" Darwin, who helped her father in his work on *The Descent of Man* (1871) (Browne 2002).

Her clear, easily readable text describes aspects of W. C. L. Martin's views on primate classification. She notes that Martin (1841:361) groups Simiadae into three subfamilies, one including the chimpanzee, the orangutan, and *Hylobates*; the second including *Semnopithecus* and probably *Colobus*; and the third including *Cercopithecus*, *Macaca*, and *Cynocephalus*. In Martin's classification, Simiadae corresponds to what we now call Catarrhini, the group including Old World monkeys and apes, except that Martin, as did others at the time, placed humans in their own order: Bimana. In *The Descent of Man*, Darwin (1871) used Simiadae to include not only Catarrhini but also Platyrrhini (New World monkeys): "The Simiadae then branched off into two great stems, the New World and Old World monkeys [including apes]; and from the latter, at a remote period, Man, the wonder and glory of the Universe, proceeded" (213).

Below Henrietta Darwin's clearly discernible text, her father notes in almost illegible pencil something concerning Ludwig Rütimeyer's ideas on New and Old

World monkeys, probably referring to his paper on Eocene Swiss mammals that also included primates (Rütimeyer 1862). Darwin also notes Jean Albert Gaudry's association of *Semnopithecus* and *Macaca*, most likely from the paper in which he describes fossils of the extinct monkey *Mesopithecus* from Greece (Gaudry 1862). There is a final comment at the very bottom about the fossil ape *Dryopithecus* from France, which was named and described by Édouard Lartet in 1856. Based on these dates, the document was written in at least 1862 but possibly before 1865, because the latter is the year in which the English biologist St. George Jackson Mivart published one of the earliest primate phylogenetic trees (see chapter 5).

Darwin's simple primate evolutionary tree on the reverse side incorporates elements from all these authors, except for Mivart's publications, which appeared later (figure 4.15). "Man," "Gibbon ~~Man~~," and "Orang ~~Gibb~~" form a four-way

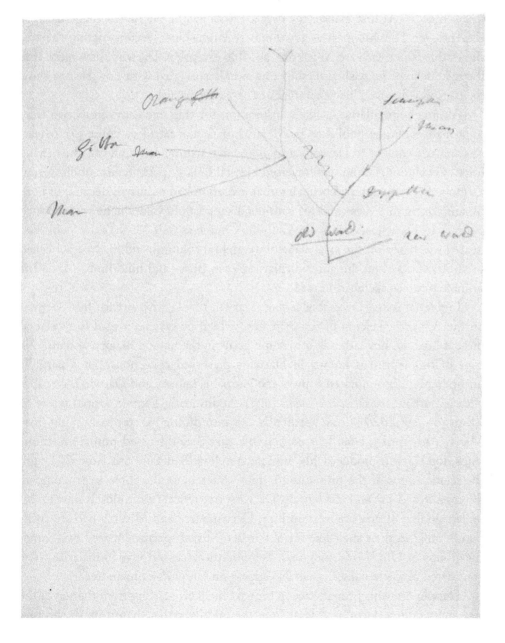

FIGURE 4.15 A simple phylogenetic tree, drawn by Darwin after 1862 and possibly before 1865 but not published (Cambridge University Library MS.DAR.80.B118v), showing general relationships among primates. (Reproduced by kind permission of the Syndics of Cambridge University Library)

split with another bifurcating branch that bears no name. This, in turn, joins a branch that bifurcates into one branch leading to "Semnopith" and the other to "Macaca." Below where these two branches join is written "Dryopithecus." Why Darwin placed *Dryopithecus* below Old World monkeys rather than below apes as described by Lartet is unclear, given that he wrote of it as being an ape about the size of a human in *The Descent of Man* (1871; Begun 2009). Below this branch Darwin wrote "Old World," with the right branch labeled "New World," both of which probably refer to the two main lineages: Old World monkeys, apes, and humans, and New World monkeys.

The second, more elaborate, and better-known primate evolutionary tree bears the date April 21, 1868 (figure 4.16). It represents the last known phylogenetic tree that Darwin produced and shows the evolutionary relationships of major groups of living primates. This is rightfully termed a rough sketch, but the roughness comes mostly from its execution and less from Darwin's uncertainty about relationships (Pietsch 2012). At first glance, the overall messiness of the sketch suggests indecisiveness on Darwin's part as to primate relationships. Deconstructing the tree shows this is not the case. With one possible exception, Darwin knew quite well the relationships he wished to show; he simply struggled with how best to show them, especially given his general lack of artistic expertise.

Although speculative, a likely inspiration for this tree came from two trees of primate evolution published by Mivart in 1865 and 1867 (see figure 5.1). Mivart was an early convert to Darwin's arguments that natural selection was the driving force of evolution, but he soon reversed himself in large part because of his devout Catholic faith, though still believing that evolution had occurred. Because of both his antiselectionist views and his continued support of evolution, he was shunned not only by Darwinists but by the Catholic Church as well. Mivart's tree published in 1865 was based on the axial skeleton (vertebral column), and his tree published in 1867 was based on the appendicular skeleton (fore- and hind limbs). Darwin's primate tree was sketched in 1868.

In a letter dated December 9, 1867, Darwin thanked Mivart for his "Memoir on the Append. skeleton of the Primates," which he says he is glad to receive at present "as I am now attending to some point in the natural history of man." No copy of this reprint is known in Darwin's preserved collections, but a partially uncut and lightly annotated copy of the journal is known, and Darwin (2005:469) cites Mivart in *The Descent of Man* (1871). Additionally, Darwin (2008) wrote to Mivart on April 6, 1868, and apparently saw him during his stay in London from March 3 to April 1, 1868. It is conceivable that they discussed primate relationships, for Darwin produced his primate tree less than a month later. Although the relationships of the primates are quite different in the three trees (compare figures 4.16 and 5.1), and incidentally Darwin's tree reflects more closely our current understanding of primate relationships, Darwin's tree and Mivart's two are strikingly similar in their stick-like form. Further, Mivart specifically used two names for groups of Old World monkeys, Semnopithecinae and Cynopithecinae (now considered obsolete names), that also appear on Darwin's primate tree.

Darwin drew his primate tree in brown ink, likely starting from the top of the tree and working down or at least from the middle, rather than from the bottom

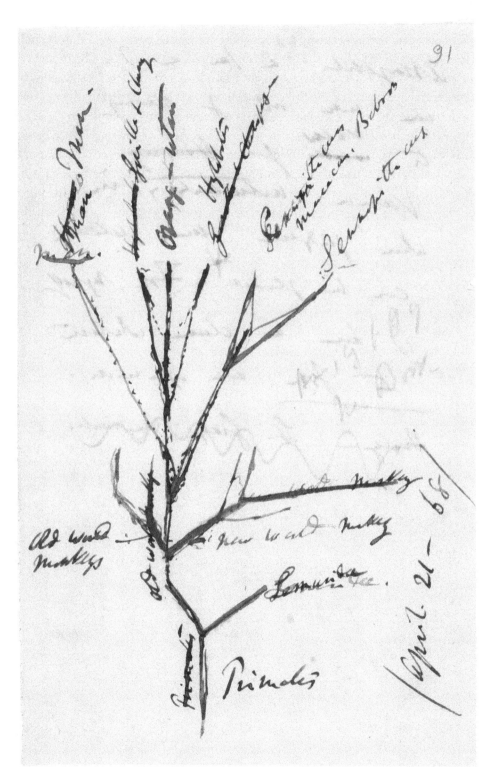

up, because the first layer of the sketch is clearly composed of dashed marks joining three main branches, which for convenience I will call the human branch on
the left, the three ape branches in the middle, and the OWm (Old World monkey) branch on the right (figure 4.17*A*). Despite the untidy juncture of the three
main branches, the human and ape branches join just above the OWm branch.
A line appears to stray from the base of human branch across to the OWm branch,
but given its closeness to the base is likely a simple slip of the pen. Dashes join the
three central ape branches as a three-way split in two places, farther up the tree and
also farther down, but the same tripartite ape occurs in both versions. On the right
for the OWm branch, the dashes form a single line. At this early stage, Darwin had
likely written "Man." horizontally above the leftmost dashed branch. Possibly, he
next inked over all three of the main branches with solid lines (see figure 4.17*B*).
On the far left "Man." was crossed out and "Homo" was written vertically in lighter
ink above it, possibly at the same time that he wrote "Hylobates" vertically on the

FIGURE 4.17 A deconstruction
and reassembly of the tree shown
in figure 4.16. The letters *A*, *B*, and *C*
refer to the three discernible stages
of construction of the tree based on
personal observation.

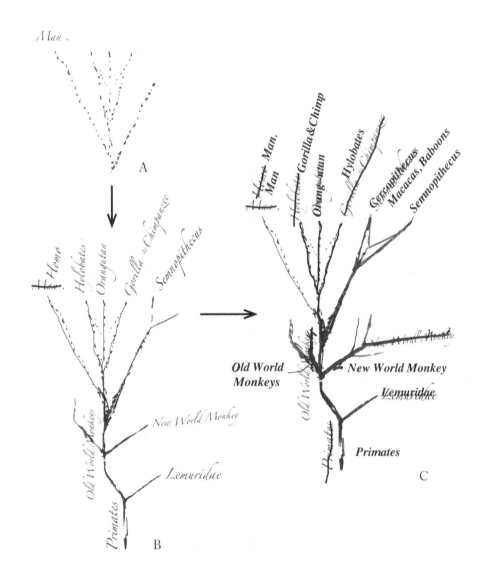

leftmost of the central branches, followed in the middle by a lightly written verti-cal "Orangutan" with a vertical "Gorilla & Chimpanzee" on the rightmost central branch. It is likely at this time that he added the two versions of bifurcating solid lines to the rightmost OWm branch and "Semnopithecus" above the leftmost of these OWm branches. Because the dashed lines do not appear to go farther down the tree, it seems that the solid ink lines in the lower part of the tree were added at this time. Darwin appears next to have written vertically "Old World Monkeys" along the main branch and probably "New World Monkey" horizontally along a right diagonal line. Below this, a right solid branch was marked "Lemuridae" in light ink, finally with "Primates" written vertically near the base. Why Darwin added the left branch labeled "Old World Monkeys" provides the one mystery about this tree. Was he simply sloppy in placing them in two places, did he see these as a group of monkeys separate from the genera of Old World monkeys he listed, or did he perceive of Old World monkeys as a larger group also giving rise to apes and humans? In the probably earlier tree in figure 4.15, he used only the terms "Old World" and "New World."

Darwin then had a change of heart about the arrangements of some names. From left to right on the top branches, he crossed out "Homo" and wrote "Man" twice; "Gorilla & Chimp" replaced a crossed-out "Hylobates"; "Orang-utan" was written over "Orangutan"; "Hylobates" replaced a scratched-out "Gorilla & Chimpanzee"; "Cercopithecus, Macacas, Baboons" is written over "Semnopithe-cus"; and "Semnopithecus" is written on the rightmost branch (see figure 4.17C). Below this, solid lines nearly obliterate "Old World Monkeys," to be replaced by "Old World Monkeys" with a line pointed at a left side branch. Four addi-tional unlabeled branches obliterate "New World Monkey," which is replaced by the same phrase below with a line pointed at this branch. Finally, "Lemuridae" is written over by the same word on the lowest right branch, and the vertical "Primates" is crossed out and replaced with a horizontal "Primates." Various lines were further darkened. By rewriting the sequence from left to right as "Man Gorilla & Chimp Orang-utan Hylobates Cercopithecus, Macacas, Baboons Semnopithecus," Darwin almost certainly thought that he showed the degree of relatedness to humans going from colobine, or leaf-eating, monkeys (*Semnopithe-cus*) on the right through to apes and humans on the left. Not readily apparent to Darwin's contemporaries was that the ordering of names in this manner does not indicate relationship; rather, the relative branching sequence indicates rela-tionship. Unfortunately this error remains quite common. Voss (2010) repeats this mistake when referring to Darwin's primate phylogeny: "The evolutionary diagram thus makes two statements about human beings: that gorillas and chimpanzees are our closest living relatives; and second, that humans represent just one part of the primate tree" (182). Whereas the first statement is not true, I agree with her second conclusion (also Pietsch 2012), but this was not, as she states, peculiar to Darwin. In 1866, for example, Ernst Haeckel published a family tree of mammals in which humans are shown as simply another branch of primates, just as Darwin did two years later in his sketch tree of primates. Very likely, Darwin knew of this tree, as he possessed a copy of Haeckel's book in his library (Cambridge University Library 1961). It is true, as Voss notes, that Haeckel (1874) also produced trees in which humans clearly sat astride the tree of life. Indeed, on the reverse side of Darwin's tree sketch in figure 4.16, he scribbled, "Arrangement as far as I can make out by comparing the ~~work~~ views of ~~Huxley~~ various naturalists as in whose judgment much reliance can be placed—For myself I have no clues whatever to form an opinion" (Cambridge University Library MS.DAR.80.B91v).

In no manner does Darwin's tree sketch elevate humans above other apes. In his diagram, Darwin correctly did group apes and humans together to the exclusion of the two main branches of cercopithecids, or Old World monkeys. Today, the hypothesis of the evolutionary branching sequence for humans and apes can be written as (gibbon(orangutan(gorilla(human + chimp)))). The parentheses circumscribe ever smaller and more terminal branches, as if one has sliced through the tree. Darwin did not know, as we do today, that chimps and humans share an ancestor, to the exclusion of other apes. He did provide the correct relative branching positions of New World monkeys and lemurs, as had other authors. As noted, the only truly incongruous branch and label is for Old World monkeys; what he meant in this instance, if he knew, remains unclear.

This evolutionary tree, which at first looks rather helter-skelter, is in fact a quite coherent exercise of how Darwin saw man's place in nature, at least as interpreted from others' work. Darwin (1871) never published any version of this tree, but as is well known, he correctly predicted the birthplace of the human lineage: "In each great region of the world the living mammals are closely related to the extinct species of the same region. It is therefore probable that Africa was formerly inhabited by extinct apes closely allied to the gorilla and chimpanzee; and as these two species are now man's nearest allies, it is somewhat more probable that our early progenitors lived on the African continent than elsewhere" (199).

Process Rather Than Pattern

What general conclusions might we draw about Darwin's various trees, all but one unpublished in his lifetime? None of the trees, except arguably the two competing about mammal origins and primate relationships, intended to show patterns of evolution for specific taxa. Rather, for Darwin the process, not the pattern, of evolution took precedence in his tree sketches. Although interested in relationships of major groups, Darwin instead used his sketches to help him work out how he perceived evolution to operate. It was not until the next generation of biologists, notably Darwin's acolyte Haeckel, that we see a clear attempt to show the pattern of evolution. These are, of course, what we most often visualize as "trees of life"— the true order of nature. As Darwin writes in *On the Origin of Species* (1859), "The affinities of all the beings of the same class have sometimes been represented by a great tree. I believe this simile largely speaks the truth" (129). Darwin was correct, and much more was to come, even in his lifetime.

CHAPTER FIVE

The Gilded Age of Evolutionary Trees

Before Darwin's *On the Origin of Species* was published in 1859, evolutionary trees of life were a novelty; after Darwin, they were a necessity, not likely because of Darwin's single tree-like diagram in this work but because of the foundations that he laid for "descent with modification by means of natural selection." The trees that ensued did not, however, bloom equally in all areas dealing with evolutionary matters. As well, for almost the next one hundred years following the establishment of a pattern of visual representation soon after Darwin, with few exceptions, we see relatively little lasting change in how the visual portrayal of evolution affected our perceptions of the process and pattern of evolution. This does not mean that this long interval of time was devoid of tree-like representations—far from it; however, as an understanding of genetics and the importance of population-based studies emerged, these trees took on new meanings.

With the turn of the twentieth century, American scientists became, for several reasons, prominent cultivators in the production and dissemination of phylogenetic trees. The science of paleontology began ascending in stature in the United States, especially in East Coast institutions. This occurred in part because of the opening of the western United States, which began in the latter part of the nineteenth century. There an incredible wealth of fossils, notably of vertebrates, sparked interest in understanding their evolutionary past. Although still a harsh environment at the time, the American West afforded far easier access than any other parts of the world with comparable fossil riches. Additionally, even though fossil invertebrates were more common, rich troves of the newly recognized dinosaurs and the best record of horse evolution, for example, ensured that depictions of vertebrate evolutionary history would predominate; and they did for the next sixty years.

The Young Turks

Darwin's reception was varied across the world, but some early, young adherents took Darwin very much to heart, producing a variety of trees related to their own scientific endeavors: the English biologist St. George Jackson Mivart (1827–1900), the French paleontologist Jean Albert Gaudry (1827–1908), the German paleontologist Franz Hilgendorf (1839–1904), the Russian paleontologist Vladimir Onufrievich Kovalevskii (1842–1883), and the German zoologist Ernst Haeckel (1834–1919) all produced evolutionary trees guided by Darwin's precepts. Uncharacteristically for a scientist, Darwin's magnum opus was published when he was already fifty years old, yet these budding scientists ranged in age from twenty to thirty-two when *On the Origin of Species* appeared, and translations into other languages took a few years. In science, as in many human endeavors, it is the younger generation that is most keen to adopt and adapt newer ideas.

One of the earliest trees to appear after the publication of *On the Origin of Species* (1859) was one by Mivart (1865), dealing with relationships of primates using features of the axial skeleton (vertebral column) (figure 5.1*A*). In 1867, Mivart published another primate tree, this time based on the appendicular skeleton (fore- and hind limbs) (see figure 5.1*B*). As Mivart (1867) points out, in the latter tree he is only trying to "express the degrees of resemblance using the appendicular skeleton of Primates, not the affinities indicated by their osteology generally, still less that evidences by the totality of their organization" (424). Both papers provide some of the earliest attempts to demonstrate that different parts of the anatomy of organisms may yield different results in understanding evolutionary relationships and in the totality of evidence needed to best understand relationships, or what we today call total evidence. Mivart must be credited with laying out very clearly the anatomical basis for his trees, but claims that he placed humans as a lateral branch on his tree (a quite modern perspective), whereas others such as Haeckel placed humans at "the apex or culmination of evolution," are incorrect (Bigoni and Barsanti 2011:6). As we will see, Haeckel, like Mivart, placed humans on different parts of their trees, depending on what was intended. Also, recall from chapter 4 that Mivart's trees likely influenced Darwin's attempts at a similarly stick-like tree that he sketched in 1868 but never published (see figure 4.15).

Mivart stands as a somewhat tragic scientific figure. Although an early acolyte of Darwin's "descent with modifications by means of natural selection," he soon turned against Darwin's ideas. In *On the Genesis of Species* (1871), although not totally rejecting natural selection, he notes that whether the theory of natural selection "be true or false, all lovers of natural science should acknowledge a deep debt of gratitude to Messrs. Darwin and [Alfred Russel] Wallace, on account of its practical utility. But utility of theory by no means implies its truth" (22). After this backhanded compliment, Mivart then compares Darwin's and Wallace's contributions with those of alchemists dealing with light emission and the atomic theories. Later he provides a list of "doubts and difficulties" with natural selection (34). Although antiselectionist, his continued argumentation in favor of evolution ran him afoul of the Catholic Church, of which he was a devote member. Both camps shunned him.

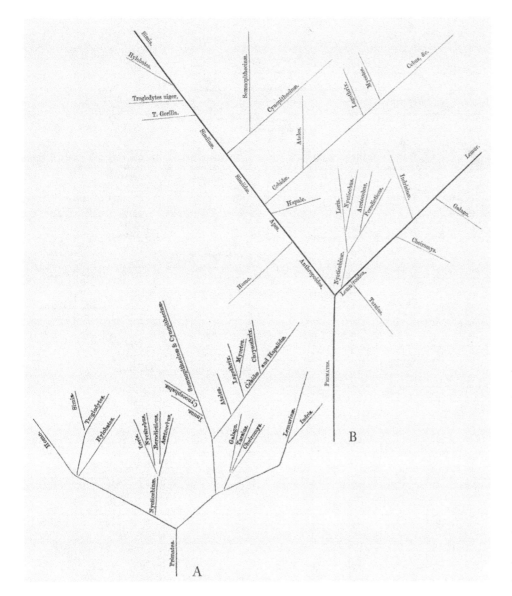

FIGURE 5.1 St. George Mivart's
trees showing relationships of
primates using features of the
(A) vertebral column, from
"Contributions Towards a More
Complete Knowledge of the Axial
Skeleton in the Primates" (1865),
and (B) fore- and hind limbs, from
"On the Appendicular Skeleton of
the Primates" (1867).

A more obscure but nevertheless interesting tree was produced by another early Darwin supporter, the German paleontologist Hilgendorf, who published in 1866 and 1867 what may have been the earliest evolutionary tree based on fossils—in this case, fossil snails from the Middle Miocene (about 14 million years ago) of the Steinheim Basin in Germany. His 1867 tree occurs as a small image in the middle of a large figure surrounded by the images of the snails he studied (figure 5.2*A*). Hilgendorf's publications in 1866 and 1867 arose from his dissertation of 1863, but this apparently did not include a tree illustration (Rasser 2006). Darwin knew of Hilgendorf, a young man in his twenties when he completed his dissertation. Beginning with the fifth edition of *On the Origin of Species* (1869), Darwin writes: "Hilgendorf has described a most curious case of ten graduated forms of Planorbis multiformis in the successive beds of a fresh-water formation in Switzerland [*sic*]" as a case showing "intermediary forms" (362). Darwin does not indicate here or in later correspondence with various people anything specifically about Hilgendorf's

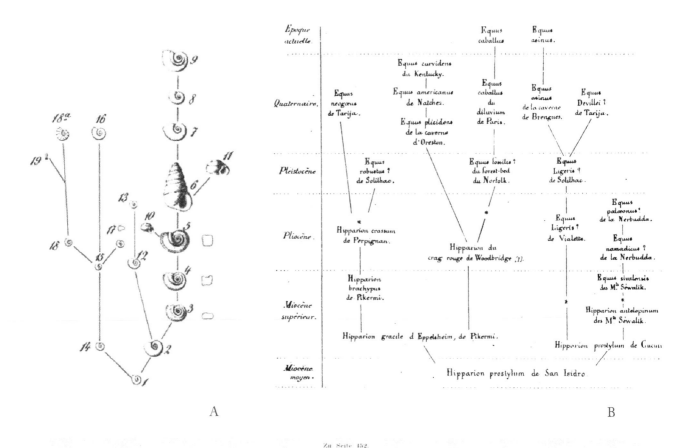

A

B

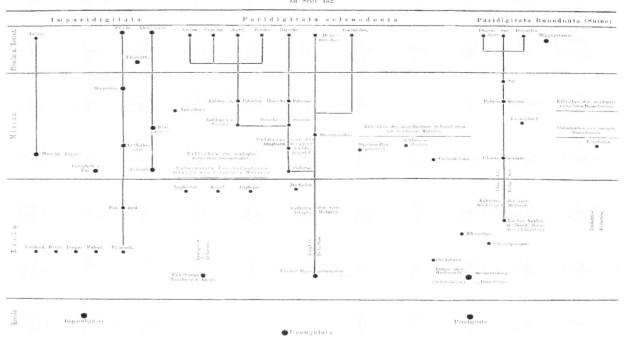

C

FIGURE 5.2 Various geologically time-based phylogenetic trees: (*A*) Middle Miocene fossil snails (about 14 million years ago) from Steinheim Basin, Germany, from Franz Hilgendorf's "Über *Planorbis multiformis* im Steinheimer Süßwasserkalk" (1867); (*B*) horses, from Jean Albert Gaudry's *Considérations générales sur les animaux fossiles de Pikermi* (1866); and (*C*) ungulates, from Vladimir Kovalevskii's "Monographie der Gattung Anthracotherium Cuv." (1876).

trees, other than noting the "graduated forms" within "successive beds" that pos-sibly allude to these trees of *Planorbis multifromis* (now *Gyraulus kleini*). Although paleontologists previously presented successions of fossils based on their fieldwork, Hilgendorf, armed with the new Darwinian theorizing, argued that this succession resulted from one species evolving into another. More recent studies show Hilgen-dorf's assessment of in situ evolution from one to three founder species within the Steinheim Basin to be more right than wrong (Nützel and Bandel 1993).

This idea that one form can give rise to another within a given area was con-troversial in Hilgendorf's day and remains so in some biological quarters. Not until more than ninety years later did this concept of evolutionary change become crys-tallized as *anagenesis*, or the change of a single population or species in one area without splitting to form two or more daughter species. The splitting of a single species or lineage to form two or more daughter species represents *cladogenesis*, or the forming of new clades. Although he did not originate the terms, George Gaylord Simpson (1961) clarified them as used today. What Hilgendorf (1867) was showing in his diagram (see figure 5.2*A*) still represents a good example of ana-genesis—in this case, in an impact crater in Steinheim Basin, Germany, in which a founder species of freshwater snail gives rise over time to a single descendant lineage of species, as well as giving rise cladogenetically to several other species (Nützel and Bandel 1993).

Gaudry (1866), one of a few French scientists who early on embraced Darwin's ideas (Tassy 2006), like Hilgendorf produced a series of paleontological trees showing anagenesis and cladogenesis in his monograph on the mammals of Pikermi, Attica, Greece. Gaudry and colleagues had worked these beds for many years, recovering thousands of bones that at the time represented thirty-five mammal genera, but Gaudry was not "content with carefully describing all these interesting forms, as were his contemporaries. In pointing out the differences, which separated them from known forms, he was led to consider the resemblances that they showed to other extinct forms or to living forms. He sought the bonds, the relationships, which united the ancient organisms to one another and to living forms" (Glangeaud 1910:422).

In this slim, sixty-eight-page monograph, Gaudry manages to present us with five quite detailed, geologically based phylogenies of hyenas, elephants and their relatives, rhinoceroses, horses and their relatives, and pigs and peccaries with their presumed ancestors and covers some 50 million years of mammalian evolution for all continents except Australia and Antarctica—quite an undertaking. Unlike Hil-gendorf, who restricted himself to arguing for the evolutionary replacement and minor splitting of one lineage of freshwater snails in a lake system in Germany, Gaudry attempted something far more grandiose: the unraveling of the evolution-ary history of five groups of mammals over great time and vast geography. Today, many phylogenetic studies based on morphology of living and extinct species, as well as molecular data, incorporate a wide a variety of species both in time and in space, as did Gaudry. Nevertheless, it now is quite uncommon to make such sweeping claims of ancestral–descent relationships within an anagenetic series of the kind drawn by Gaudry, whereas the much more geographically and usually more time-restricted kind of studies done by Hilgendorf are found today. It must be emphasized that neither of these studies harkens back to the ladders or *scala naturae* found in the works of Charles Bonnet or even in Jean-Baptiste Lamarck's earlier endeavors, which imagined a step-like or even seamless chain of being. Rather, both Hilgendorf's limited and Gaudry's expansive studies are attempts at unraveling what these authors hypothesized as very specific ancestor–descendant relationships. Gaudry's tree for horse evolution appears in figure 5.2*B*. As discussed in chapter 1, phylogenies of horses as well as that for groups such as elephants and their relatives served as part of the backdrop for the first half of the twentieth century as to how evolution occurs and how it should be drawn.

The fourth and youngest of these aspiring scientists was the Russian paleontologist Kovalevskii, who translated into Russian the works of a number of well-known European scientists, including some of the works of Darwin. In search of a research topic to further the Darwinian view of evolution, Kovalevskii was influenced by Thomas Henry Huxley's interest in the evolution of horses and other ungulate, or hooved, mammals. Kovalevskii's (1876) major work in this area resulted in a well-argued, three-part monograph completed in late 1873 through early 1874 that Gaudry highly praised (Vucinich 1889). Figure 5.2*C* shows Kovalevskii's quite rectilinear trees, the style of which became popular in the 1970s to indicate Niles Eldredge and Stephen Jay Gould's (1972) punctuated equilibrium hypothesis of evolutionary rates, which emphasized rapid speciation followed by stasis. Even if he appears prescient, this was not Kovalevskii's intent; it simply was his manner of drawing. The scale is too small to make out details, but on the left the three vertical lines show his anagenetic evolutionary view for tapirs, horses, and rhinos (what today we call perissodactyls, or odd-toed ungulates). The most complex tree, in the middle, is for cattle, sheep, antelope, and the like (what today we call ruminant artiodactyls, or even-toed cud chewers). On the far right, the tree is for true pigs and peccaries, or suiforms. The hippopotamus is shown as a dot at the top right but is not connected to the suiform tree. It is now known to be the nearest living relative to all cetaceans and not closely related to suiforms. There is an abbreviated geological scale on the left starting back in the Cretaceous period, over 65 million years ago, and ending at the present day. At the bottom middle is a large dot for the imagined Urungulata, or protohooved ancestor. Interestingly, we now know with some certainty that the earliest known ungulate, aptly named *Protungulatum*, is indeed a rare occurrence in the Cretaceous of North America (Archibald et al. 2011). Of considerable interest in Kovalevskii's figure is what he does not attempt to show; he leaves many taxa as dots not connected to other dots. He does group them in a general way with the other taxa that he believes they might be nearest to evolutionarily, but he sounds a cautionary note about what he thinks we do not know in leaving dots unconnected. If only some modern systematists were so cautious in drawing relationships.

Haeckel, the last of the young Turks discussed here, early on set out to test and promote Darwin's theories. A brilliant and controversial German biologist, Haeckel earned the sobriquet "German Darwin." He produced a veritable thicket of quite different sorts of trees and in so doing dominated this early phase of the gilded age of evolutionary trees. Haeckel coined a number of the biological terms still used today, the most germane for this discussion being "ontogeny" for individual organismic development and "phylogeny" for evolutionary development—hence the now common use of phrases such as "phylogenetic tree" (of obvious meaning) and "phylogenetic systematics" for the study of the evolutionary relationships of organisms.

Haeckel's Thicket of Trees

Ernst Haeckel's inspiration to organize and picture life in some sort of natural system began in the summer of 1860 with his first encounter with Heinrich Bronn's German translation of and commentary on Darwin's *On the Origin of Species*. At the time, Haeckel was in his mid-twenties and was attempting to complete his *Habilitationsschrift*, his postdoctoral work on radiolarians, a kind of protozoan that produces a mineralized skeleton that rains onto the ocean floor, forming radiolarian ooze. This work was published as an impressive two-volume monograph in 1862. But in the fall of 1861, while finishing his postdoctoral research, Haeckel once again became immersed in *On the Origin of Species*, which profoundly affected his scientific efforts for the remainder of his life (Richards 2008).

Starting with the beautiful illustrations that Haeckel (1862) produced for his monograph on radiolarians, it was clear that few people possessed his combination of artistic ability and scientific knowledge to place evolutionary history and theory in the form of wondrous, tree-like diagrams, or phylogenetic trees. It is no exaggeration to say that in the realm of producing exquisite illustrations of natural history, he ranks with the likes of John J. Audubon in the United States and John and Elizabeth Gould in England.

In 1866, in the second volume of his two-volume work *Generelle Morphologie der Organismen* (*General Morphology of Organisms*), which, unfortunately, was not widely read or translated, Haeckel prepared no fewer than eight phylogenetic trees. His first tree addresses adroitly one of the problems that vexed not only Darwin but also most biologists in the 1860s, and in fact still does: How did life begin, and how many times did it do so? Recall that Lamarck wrote of multiple origins and that Carl Edward von Eichwald's many-trunked tree indicated the same idea (see figure 3.3). The translated title of Haeckel's tree in the lower-right corner reads, "Monophyletic Family (or Genealogical) Tree of Organisms" (figure 5.3). In this volume, Haeckel also coined the word "Moneren" for what we term informally today as monerans for simple, single-celled organisms, but which we no longer use in formal taxonomy.

The lower-left corner in figure 5.3 lays out his three hypotheses. Starting with the lowest, "III, box: pstq (1 branch)," bounded by the letters p, s, t, and q, this hypothesis shows a single or monophyletic origin of all life from what Haeckel calls in Latinized phrasing "a single self-seeding common root of organisms." From the single branch, left to right are the kingdoms "Plantae," "Protista," and "Animalia." The second hypothesis, "II, box: pxyq (3 branches)," argues that plants, protistans, and animals arose separately. Finally, hypothesis 3, "I, box: pmnq (19 branches)," argues that six plant, eight prostistan, and five animal groups or lineages originated separately. These nineteen branches are labeled throughout the tree as well as along the mn line that crosses the figure.

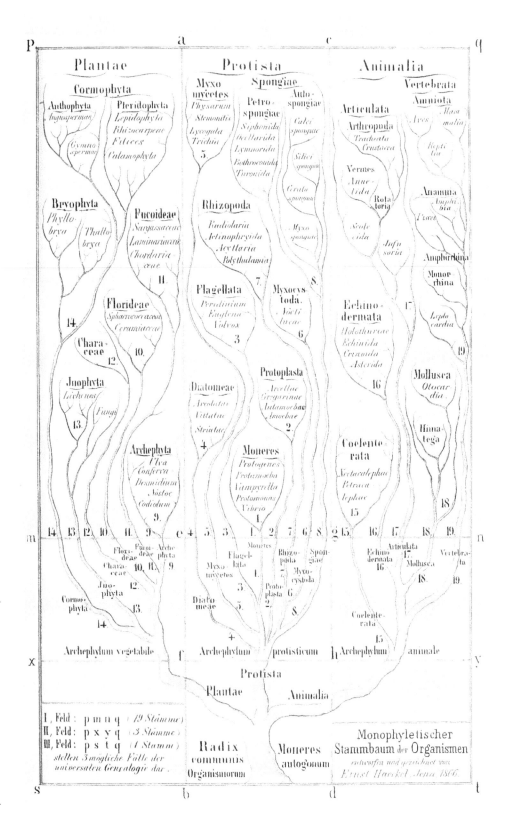

FIGURE 5.3 Ernst Haeckel's "Monophyletic Family (or Genealogical) Tree of Organisms," showing his three hypotheses for the origin of life: a single moneran gave rise to all life; three different monerans originated and gave rise to plants, other monerans, and animals; or multiple lineages arose from many original monerans, from *Generelle Morphologie der Organismen* (1866).

With these three hypotheses from Haeckel's first tree in *Generelle Morphologie*, he hedged his bets on the origin of life: a single moneran gave rise to all life; three different monerans originated and gave rise to plants, other monerans, and animals; or multiple lineages arose from of many original monerans (Richards 2008). In this work Haeckel, vacillates on which hypothesis he supports, but possibly because Darwin favored the third hypothesis in *On the Origin of Species*, Haeckel did as well.

Theodore Pietsch (2012) reproduces in considerable detail this and the remaining seven trees in Haeckel's *Generelle Morphologie* in a visual compendium of trees of life, but the latter seven are reproduced in figure 5.4 at a much-reduced scale to simply convey their general form. The Roman numerals indicate Haeckel's numbering scheme. All eight trees present the same general appearance of wispy, somewhat gnarled, almost vine-like forms with multiple or thin main stems. In a number of instances, the relationships shown no longer pertain, but as noted, it is their form that is of interest here. All seven trees show relative branchings and hence relations of one group to another. All seven trees also include extinct forms known only as fossils, and, as far as I can determine, none are shown as ancestors but rather as branches on the tree. When a name appears across a single major branch, the intent is to indicate what Haeckel perceives as the relative grade of evolution on the tree at that point in relative evolutionary time. Most such taxa then repeat across a number of smaller branches farther along the tree or appear terminally, with brackets superintending the groups they include. For example, in the uppermost left phylogeny of plants in figure 5.4, Gymnospermae is found both as a grade (*lower arrow*), leading to both gymnosperms and angiosperms, and as a higher taxon written across extant branches (*upper arrow*). This explains Haeckel's reference to more primitive and more advanced grades in evolutionary history—shades of Aristotelian ladders along with his tree motif.

Five of the seven trees have an additional tree in a lower corner showing the same relationships portrayed on the larger tree, but with fewer names to help the reader understand higher-level relationships without the clutter of additional smaller group names. The one significant difference is that the trees for echinoderms and vertebrates are, as Haeckel notes, "paleontologically based," meaning that he has provided a geological scale on the left side of each of these figures into which extinct forms and the origin of groups are placed at the appropriate geological interval, but again ancestors are not identified. In all eight figures and all the included trees, he shows us the relative branching of taxa, but not any ancestor–descendant relationships.

The importance of real or hypothetical ancestors comes with Haeckel's next major set of trees beginning with *Natürliche Schöpfungsgeschichte* (*The History of Creation*, 1868). He based this work on a series of public lectures on the theory of evolution and how it applies to human origins, which helped bring Haeckel to prominence as one of the major proponents of evolutionary theory at the time. Unlike *Generelle Morphologie der Organismen* (1866), this newer work was more widely read and translated, hence his broader renown. In it, he provides us with fourteen figures that he identifies as *Stammbaum* (*Stammbäume*), a family or genealogical tree(s). One of these, labeled "Ahnenreihe des menschlichen Stammbaums"

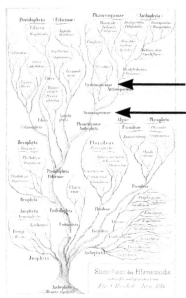

II Plants

III Coelenterates

V Articulates

VI Mollusks

VIII Mammals

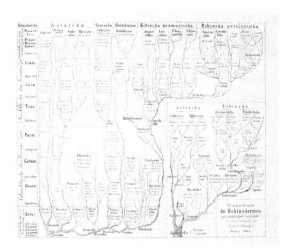

IV Echinoderms

VII Vertebrates

(Line of Ancestors of the Human Family Tree), provides no tree, but two columns. The two left columns indicate geologic time. Translations of the two right columns are "Animal Stages of Human Ancestors," indicating relatively higher taxonomic groups, and "Living Closest Ancestral Stages," indicating supposed living examples of the respective groups—a ladder-like array showing relative order of appearance or ascendance during geologic time, ending at the top with peoples whom Haeckel and many others at the time regarded as primitive humans.

Of the other thirteen at least partially branching diagrams, four forms are represented: five trees use a bracketed architecture showing more inclusive groups as one moves up the tree; two have geologic time scales and are even wispier versions of those in *Generelle Morphologie*, resembling interconnected strands of upside-down feathery Spanish moss; five have broom-like tufts, one emanating from the next lower; and one has a geologic time scale but in very severe rectilinear form. Figures 5.5 and 5.6 show examples of these four kinds of tree, with more examples available in Pietsch's (2012) tree diagram compendium.

Three of the five trees with the bracketed architecture examine ideas on the polyphyletic, or multiple-origin, hypotheses of life showing various amounts of branching. The other two, decidedly more branching, detail the evolutionary history of hooved mammals in one and that of primates in the other. The diagrams also include some real and imagined ancestral forms. Figure 5.5*A* is one of the kinds of these bracketing diagrams—in this case, dealing with the origin of plants. At the base of the diagram Haeckel indicates "numerous vegetable monerans, created independently by spontaneous generation," leading separately upward to ferns, lichens, and so on, as well as to other groups that did not lead anywhere, as indicated by the question marks. In the middle, he shows a ladder-like ascent starting with the Archephyta (ur-plant) near the bottom up to the angiosperms (*Decksemige*) at the top. Haeckel notes in the explanation of this figure that in plate II, he shows the alternative view of plant origins as monophyletic (see figure 5.5*B*). The plant taxa shown along the top are essentially the same as those shown in figure 5.5*A*, and along the left is a geologic time scale, so that here we see Haeckel's view as to when various major groups of plants arose. Excepting some question marks on the figure, no text appears on the phylogeny. Figure 5.5*B*, along with a similar one for animals, represents among the most finely grained branching diagrams ever produced. The number of branchings at any given geologic interval varies considerably, so the relative density represents how speciose any given branch is at any given time. For example, Haeckel speculates (indicated by question marks) that angiosperms appear before the Triassic period, but the increase in density of branchings shows a great increase in species numbers beginning in the Cretaceous period. The small inset tree in figure 5.5*B* is reminiscent of Haeckel's trees from 1866 in its wispy, vine-like form, which shows simplified, higher-level groupings of plants.

FIGURE 5.4 Haeckel's seven phylogenetic trees for the evolution of plants and major groups of animals, from *Generelle Morphologie der Organismen* (1866).

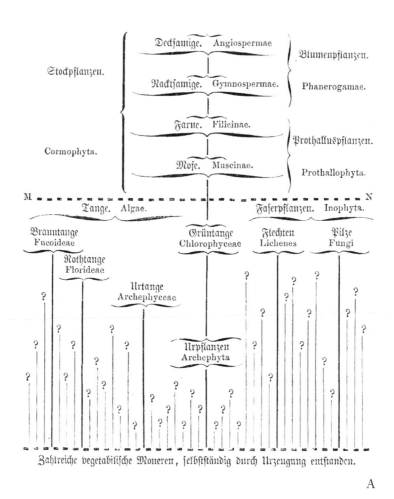

A

B

Figure 5.6*A* shows one of Haeckel's (1868) five broom-like tuft diagrams, chosen for representation here because it repeats in a slightly different form the "Family Tree of Organisms" from *Generelle Morphologie* (1866) (see figure 5.3), which also addressed the issue of the poly- versus monophyletic origin of life. Beyond the obvious differences in the wispy, vine-like versus the broom-like tufts, the figures are similar and different in some other important aspects. They are similar in showing the three versions of monophyletic, triphyletic, and polyphyletic origin of life—even using the same letters to categorize each level. They are different in several ways, all intended to pander to a wider audience—recall that *Natürliche Schöpfungsgeschichte* (1868) sprang from popular lectures and was far more widely read and translated than *Generelle Morphologie* (1866). This shows in his trees dealing with the origin of life. In the figure from 1866, he is first and foremost concerned, after the question of monophyly versus polyphyly, with showing the branching patterns of the main groups within plants, prostistans, and animals, identifying them with Latinized names. In the figure from 1868, he adds German words as well as the Latinized words for public consumption, but more important is the concept of ancestry. He indicates an "Urstamm," or archetypal stem, for plants, protistans, and animals. Above this are noted "Ur-plants," "Ur-beings," and "Ur-animals." Above this, we find some branching, but the idea of a ladder-like form has taken hold. Especially compare the "plants" on the left sides in figures 5.3 and 5.6. In the former, the relationships are largely of a branching nature, whereas in the latter, they for the most part form a ladder: green algae, moss, ferns, Gymnosperms, angiosperms. The idea of an "ur-," or archetypal, organism comes from the antecedent of science, natural philosophy, as embodied in Goethe's idea of "ur-phenomenon" (*Urphänomen*), which Haeckel cannot quite escape (Seamon 1998). His Darwinian side pushed for the branching of life, whereas his intellectual upbringing in natural philosophy pulled him toward Goethe's ideas of the "Ur-" that still enticed the general populace, helping to create the popularity of his 1868 book.

As a coda to his trees in *Natürliche Schöpfungsgeschichte*, I show Haeckel's single severe, rectilinear tree (see figure 5.6*B*). Although produced in the 1860s, its design would have been equally comfortable in the 1920s and 1930s Bauhaus tradition. It again identifies "Ur-" taxa, here leading to six major lineages of animals placed within a geologic time scale. One could read what we today term parallel evolution within a group because of the nearly parallel lines found in his trapezoidal shapes, but this would be hyperbole. In the caption, he writes that one should read the text for explanation, but reading the text tells us nothing of why he constructed the phylogeny in the manner he did. It remains a mystery, other than suggesting that between his 1866 work, in which all the figures had a similar form, and his 1868 work, for which he gained much more popular

FIGURE 5.5 One of Haeckel's (*A*) trees of plant evolution indicating multiple origins by means of bracketed architecture, with more inclusive groups as one moves up the tree, and (*B*) wispier versions of plant evolution indicating a monophyletic origin, from *Generelle Morphologie der Organismen* (1866).

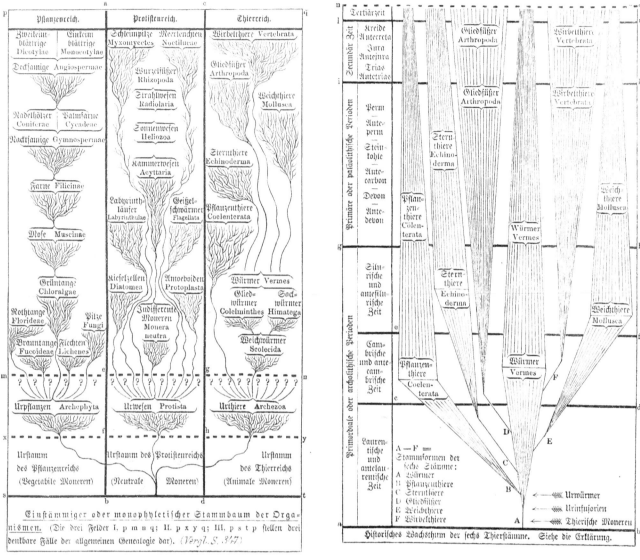

A

B

FIGURE 5.6 One of Haeckel's (A) broom-like tuft diagrams, addressing the issue of the poly- versus monophyletic origin of life, depicted in a slightly different form from his figure in *Generelle Morphologie der Organismen* (see figure 5.3), and (B) single severe, rectilinear tree of animal evolution, placed in a geologic time scale, from *Natürliche Schöpfungsgeschichte* (1868).

acclaim, including as a consummate artist, he decided to play with different phylogenetic forms. The only real scientific change was the admittance of ancestors into the phylogenetic fold—not a new idea, but one that seems to have haunted Haeckel's later work, especially pertaining to human evolution and his views on human races (and even different human species [see chapter 8]). In later editions of *Natürliche Schöpfungsgeschichte*, in both the German (1873) and the English and American editions (1876), Haeckel used a wispier derivative of a very similar tree. This reinforced the idea that he simply was playing with design and representation in this tree; the severe rectilinear form disappeared.

Haeckel must be given credit for another innovation in his use of tree-like figures that first appeared in *Natürliche Schöpfungsgeschichte*, but not until the second edition, published in 1870. Scientists such as Alexander von Humboldt (1817) had produced altitudinal zonations of vegetation; others such as Théodore Lacordaire (1839–1840) had drawn up tables indicating the global distributions of various

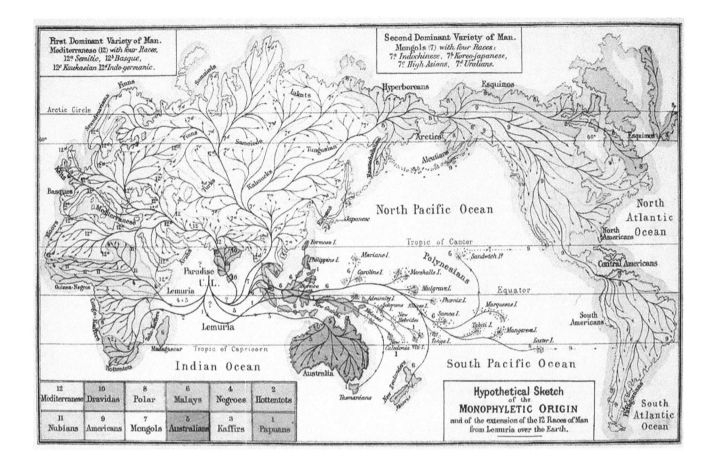

insects (Browne 1983); and still others such as Alfred Russel Wallace (1876) deserve credit for helping to establish the science of biogeography, which deals with the geographic distribution of plants and animals. In the second edition of *Natürliche Schöpfungsgeschichte*, Haeckel published what appears to be the first map on which is superimposed a tree-like form, albeit a very complex form, that spreads over its surface and shows the origin and radiation of a group—in this case, of humans (figure 5.7). Although Haeckel had no knowledge of genetics, clearly his contribution is an antecedent of today's study of phylogeography, which concerns the biogeographic spread and distribution of populations, mostly using genetic data. He places the origin of humans on a hypothetical continent in the Indian Ocean. Named Lemuria (presumably after *lemur*, Latin for "ghost of the departed") in 1864 by the English zoologist Philip Sclater, it supposedly connected various regions in the Indian Ocean, including Madagascar, home to real lemurs. Haeckel explains in the text that Lemuria subsequently disappeared below the waves. Although quite fanciful sounding to us, the idea of lost continents as stepping-stones for the spread of plants and animals was not an uncommon notion until the theory of continental drift took hold after the middle of the twentieth century, rendering the need for such imagined lost continents generally unnecessary.

The final work of Haeckel discussed here, *Anthropogenie; oder, Entwickelungsgeschichte des Menschen* (*Anthropogeny; or, Evolutionary History of Man*, 1874), like *Natürliche Schöpfungsgeschichte* (1868), proved popular enough to be translated

A

Uebersicht über das phylogenetische System der Säugethiere.

I. Erste Unterklasse der Säugethiere	Kloakenthiere (**Monotrema** oder **Ornithodelphia**)	1. Stammsäuger — *Promammalia*	
		2. Schnabelthiere — *Ornithostoma*	
II. Zweite Unterklasse der Säugethiere	Beutelthiere (**Marsupialia** oder **Didelphia**)	3. Pflanzenfressende Beutelthiere — *Botanophaga*	
		4. Fleischfressende Beutelthiere — *Zoophaga*	

III. Dritte Unterklasse der Säugethiere **Placentalthiere** (**Placentalia** oder **Monodelphia**)

III A. Placentalthiere ohne Decidua, mit Zotten-Placenta **Indecidua Villiplacentalia**
- 5. Hufthiere **Ungulata** — Unpaarhufer *Perissodactyla* / Paarhufer *Artiodactyla*
- 6. Wallthiere **Cetacea** — Seerinder *Sirenia* / Wallfische *Sarcoceta*
- 7. Scharrthiere **Effodientia** — Ameisenfresser *Vermilinguia* / Gürtelthiere *Cingulata*

III B. Placentalthiere mit Decidua, mit Gürtel-Placenta **Deciduata Zonoplacentalia**
- 8. Scheinhufthiere **Chelophora** — Klippdasse *Lamnungia* / Elephanten *Proboscidea*
- 9. Raubthiere **Carnassia** — Landraubthiere *Carnivora* / Seeraubthiere *Pinnipedia*
- 10. Halbaffen **Prosimiae** — Fingerthiere *Leptodactyla* / Langfüsser *Macrotarsi* / Pelzflatterer *Ptenopleura* / Lemuren *Brachytarsi*

III C. Placentalthiere mit Decidua, mit Scheiben-Placenta **Deciduata Discoplacentalia**
- 11. Nagethiere **Rodentia** — Eichhornartige *Sciuromorpha* / Mäuseartige *Myomorpha* / Stachelschwein-artige *Hystrichomorpha* / Hasenartige *Lagomorpha*
- 12. Insectenfresser **Insectivora** — Blinddarmträger *Menotyphla* / Blinddarmlose *Lipotyphla*
- 13. Flederthiere **Chiroptera** — Flederhunde *Pterocynes* / Fledermäuse *Nycterides*
- 14. Affen **Simiae** — Plattnasen *Platyrhinae* / Schmalnasen *Catarhinae*

B

Stammbaum der Säugethiere.

Menschen Homines — Elephanten Proboscidea — Menschenaffen Anthropoides — Fledermäuse Nycterides — Klippdasse Lamnungia — Seeraubthiere Pinnipedia — Scheinhufer Chelophora — Schmalnasen Catarhinae — Flederhunde Pterocynes — Plattnasen Platyrhinae — **Flederthiere Chiroptera** — Landraubthiere Carnivora **Raubthiere Carnassia** — **Nagethiere Rodentia** — **Affen Simiae** — Pelzflatterer Ptenopleura — Waldische Sarcoceta — Fingerthiere Leptodactyla — Lemuren Brachytarsi — Insectenfresser Insectivora — Langfüsser Macrotarsi — Seerinder Sirenia **Walthiere Cetacea** — Faulthiere Bradypoda — **Hufthiere Ungulata** — Scharrthiere Effodientia — Halbaffen Prosimiae **Deciduathiere Deciduata** — **Decidualose Indecidua** — **Placentalthiere Placentalia** — Pflanzenfressende Beutelthiere Marsupialia botanophaga — Fleischfressende Beutelthiere Marsupialia zoophaga — Schnabelthiere Ornithostoma — **Beutelthiere Marsupialia** — Stammsäuger Promammalia **Kloakenthiere Monotrema**

FIGURE 5.8 Haeckel's (*A*) "Overview of the Phylogenetic System of Mammals," which to my knowledge is the first classification identified as being phylogenetic, and (*B*) accompanying "Genealogical Tree of Mammals," from *Anthropogenie; oder, Entwickelungsgeschichte des Menschen* (1874).

into English. In it, he presents two "genealogical trees" (*Stammbäume*), one for vertebrates (*Wirbeltheire*) and one for mammals (*Säugethiere*), each of which is accompanied by a classification that Haeckel calls "Uebersicht über das phylogenetische System" (Overview of the Phylogenetic System) for each particular group. Figure 5.8*A* shows the phylogenetic classification of mammals, and figure 5.8*B* displays the accompanying "Genealogical Tree of Mammals." To my knowledge, these are the first classifications identified as phylogenetic, but this is not surprising inasmuch as Haeckel coined the term "phylogeny."

Today, for a biologist doing systematics, the modifier "phylogenetic" indicates a one-to-one correlation between the classification and the phylogeny, such that one can produce the classification based on the phylogeny and vice versa. Such is not the case for Haeckel's trees and classifications, because in his trees, Haeckel shows not only branching of various groups, which in principle can be indicated in the classification, but also progression from what he perceives as more primitive to more advanced groups. The classification thus cannot be constructed from the tree topology and thus in today's usage would not be considered phylogenetic, as it does not reflect the evolutionary history of the group. For example, on the left side of the tree we see ungulates (hooved mammals) giving rise to cetaceans (whales and relatives) (see figure 5.8*B*), whereas in the classification, they are coeval groups (see figure 5.8*A*). Arguably, truly phylogenetic classifications along with their accompanying trees did not appear with any frequency until almost one hundred years later, but again it was a German who developed these newer schemes of discerning evolutionary relationships.

This brings us to Haeckel's most famous tree in *Anthropogenie*, which in some ways represents the granddaddy of all phylogenetic trees, also combining progress with branching evolution (figure 5.9). This tree is titled "Stammbaum des Menschen" (Genealogical Tree of Humans). This is a massive specimen, so much so that it sometimes is referred to as Haeckel's "great oak" (Pietsch 2012). Its gnarled, impressive form would make Edward Gorey proud. Read simply, it shows the ascendancy from the primordial moneran to the pinnacle of evolution— "Menschen," or humans. Insects, which we today surmise number in the tens of millions of species, are a small side branch on the left of the oak. Haeckel did not know the true magnitude of this number, but I have little doubt that he realized that insect species far outnumbered mammal species.

Ink has been spilled interpreting this tree, arguing whether such a more ladder-like representation was somehow non-Darwinian (Bowler 1988; Dayrat 2003) or was simply attempting to show human evolutionary history from one perspective, whereas Haeckel's other trees show humans as a mere branch on the tree of life (Richards 2008). I am inclined to the latter view, especially because as shown, Haeckel varied how he represented the evolution of other groups, as twigs or as central trunk. This said, Haeckel's oak is different not merely in its massive appearance but also in its intended or unintended representation. Like Darwin, Haeckel was disinclined to reach for teleological, final, deity-based causes; yet like Darwin, Haeckel did see progress in evolution. Thus later species were more evolved or higher on the scale of life because of natural selection (Richards 2008). Unlike on the trees noted earlier on which higher taxa appear twice, first as a grade (or level organization) and then defining a clade (or branch) (for example, Gymnospermae in figure 5.4), in Haeckel's oak higher groups appear generally only as grades where they fall along the progress toward humans. Thus this tree is certainly progressive, yet need not be teleological. Most likely, there is no ulterior motive other than Haeckel, as a brilliant scientist and equally talented artist, playing to an audience who could handle Darwin's evolution as long as they themselves crowned the tree of life. Humans and especially Europeans were not yet prepared to see themselves as just one more evolutionary twig. Haeckel advanced this view with blatantly racist ideas of human evolution and dispersal.

Waning of an Age

Haeckel's trees dominated the iconography of evolutionary relationships in the late nineteenth century if for no other reason than he produced so many tree-like phylogenies. A number of other tree-like as well as untree-like figures were

Stammbaum des Menschen.

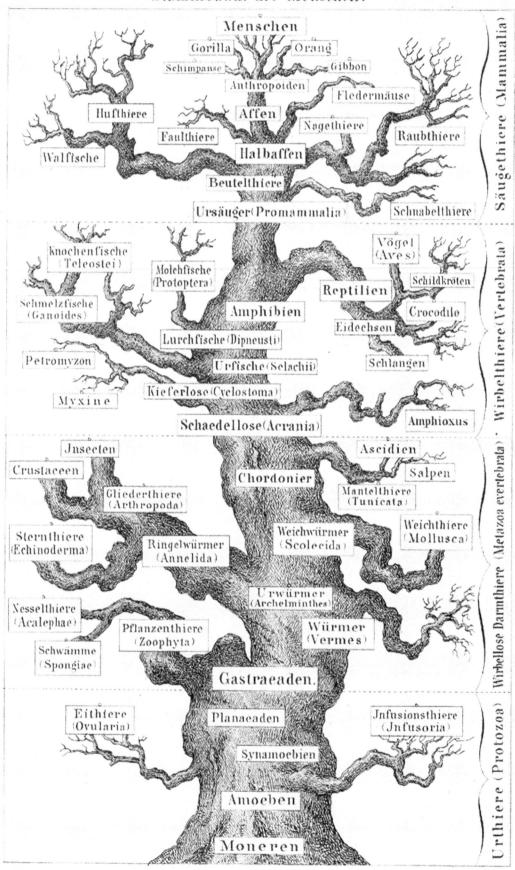

Menschen

Gorilla — Orang

Schimpanse — Gibbon

Anthropoiden

Affen — Fledermäuse

Hufthiere — Nagethiere

Faulthiere — Raubthiere

Walfische — Halbaffen

Beutelthiere

Ursäuger (Promammalia) — Schnabelthiere

Säugethiere (Mammalia)

knochenfische (Teleostei) — Vögel (Aves)

Molchfische (Protoptera) — Schildkröten

Reptilien

Schmelzfische (Ganoides) — Crocodile — Amphibien — Eidechsen

Lurchfische (Dipneusti)

Petromyzon — Urfische (Selachii) — Schlangen

Myxine — Kieferlose (Cyclostoma)

Schaedellose (Acrania) — Amphioxus

Wirbelthiere (Vertebrata)

Jnsecten — Ascidien

Crustaceen — Salpen

Chordonier

Gliederthiere (Arthropoda) — Mantelthiere (Tunicata)

Sternthiere (Echinoderma) — Ringelwürmer (Annelida) — Weichwürmer (Scolecida) — Weichthiere (Mollusca)

Urwürmer (Archelminthes)

Nesselthiere (Acalephae) — Würmer (Vermes)

Pflanzenthiere (Zoophyta)

Schwämme (Spongiae)

Gastraeaden.

Wirbellose Darmthiere (Metazoa evertebrata) · Wirbellose Darmthiere

Eithiere (Ovularia) — Planaeaden — Jnfusionsthiere (Jnfusoria)

Synamoebien

Amoeben

Moneren

Urthiere (Protozoa)

published during this time, but none offered truly new ideas of visual representation that found much of an audience. Pietsch (2012) provides a nice representation of untree-like and tree-like figures from the late nineteenth century. Among the former, George Bentham (1800–1884) shows in a diagram published in 1873 the group of flowering plants known as the composites as a series of interconnected circles, whereas in 1896 Nikolai Ivanovich Kuznetsov (1864–1932) presents a decidedly untree-like network of interrelationships of the flowering plant *Gentiana*. Then there is the somewhat more tree-like chart of the animal kingdom encased in a rounded cone published by Graceanna Lewis (1821–1912) in 1868, which may be only the second such figure by a woman, Anna Maria Redfield's (1857) being the first (see figure 3.12).

The most ingenious, innovative, and complex yet largely forgotten trees produced at the time belong to Max Fürbinger (1846–1920). Fürbinger's formal studies began in 1865 at Jena, Germany, and he completed his dissertation in Berlin on the skeleton and musculature of lizards with reduced limbs (*Die Knochen und Muskeln der Extremitäten bei den schlangenähnlichen Sauriern*, 1870). Although he worked widely in vertebrates, his research centered on reptiles and birds, and in *Untersuchungen zur Morphologie und Systematik der Vögel, zugleich ein Beitrag zur Anatomie der Stütz- und Bewegungsorgane* (*Studies on the Morphology and Systematics of Birds, also a Contribution to the Anatomy of the Locomotor Organs*, 1888), we find the quite amazing phylogenetic trees of extinct and living birds. The figures include two very wispy trees titled "Experimental Family Trees of Birds" and three "Horizontal (Planimetric) Projections of the Family Trees of Birds."

What at first glance appear to be mirror-image trees actually show views of the same tree from either side, with various groups of birds emphasized from each view (figure 5.10*A* and *B*). The arrows indicate lines that Fürbinger drew across the trees, dividing them into lower, middle, and upper zones (figure 5.10*C–E*), which are shown in the three horizontal projections. These do not show exact slices through the tree at one given level but compress top–down or bottom–up views of the tree within each of the three zones, with distances between circles suggesting the degree of divergence. The circles and enclosing outlines are named for the relevant bird groups and the number of species in the particular groups. The size of a circle also indicates the relative number of species in that group.

These diagrams represent the first attempt at presenting a three-dimensional phylogenetic tree, albeit limited by nineteenth-century technology. Fürbinger (1888:1751) terms these diagrams his "Genealogical System" and explains the difficulty of representing the relationship as a "stereometric family tree," resulting in the two trees and three cross sections. The attempt at three dimensionality is clear when he explains that in "the vertical views the front branches are drawn sharper and stronger, the rear expressed finer and weaker" and that in the horizontal views the deeper and larger branches are represented by fine dotted lines. On an earlier

FIGURE 5.9 Haeckel's "Genealogical Tree of Humans," from *Anthropogenie; oder, Entwickelungsgeschichte des Menschen* (1874).

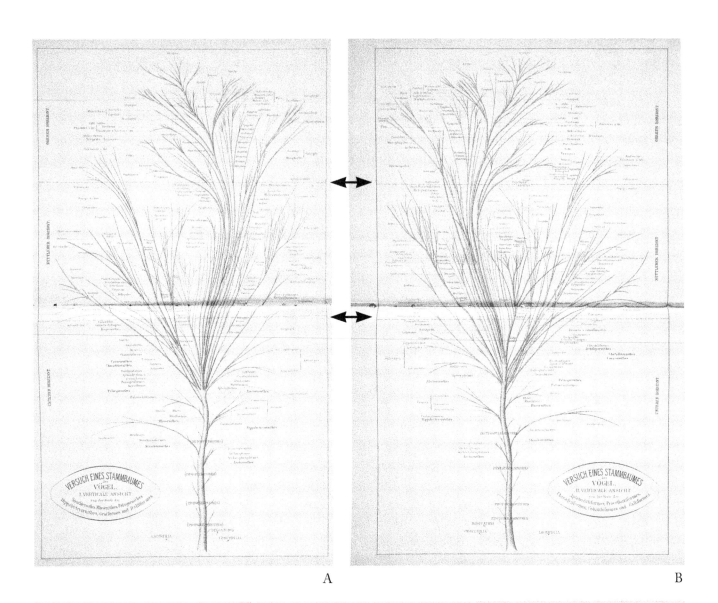

A

B

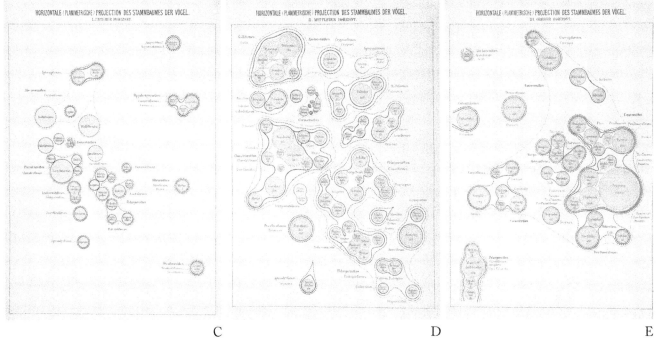

C   D   E

page, Fürbinger suggests that he could produce a "three-dimensional family tree" by "custom building it out of wire or a similar material" (1569), which I assume he never built.

The base of the tree shows birds arising from a common ancestor with dinosaurs, crocodilians, and lizards and relatives. Up the main stem follow evolutionary grades of birds. In the late nineteenth century, Thomas Henry Huxley suggested that birds had descended from dinosaurs but did not defend this idea in the face of criticism. The idea was circulating among other scientists, though, and Fürbinger (1888) makes considerable comparison between birds and dinosaur anatomy, even citing other authors who said that the oldest known bird "*Archaeopteryx* and other flying birds and with them the ratites [flightless ostriches and relatives] developed through dinosaurs" (1142). In the end, as his phylogenetic tree shows, Fürbinger could not decide with certainty from which groups of extinct or extant reptiles the birds sprang.

Separate from any issues about the correctness of the phylogenetic relationships shown, Fürbinger's trees and cross sections are, unfortunately, far too complex to be easily read and comprehended even though each tree is quite large, spreading across two pages with each horizontal projection a full page. Where Haeckel used unnamed multifarious branchings in his trees to simply indicate taxonomic richness of a group, Fürbinger tried to name almost every wispy branch of his trees. In the end, it was simply too much for the reader to grasp. This was an innovation before its time.

FIGURE 5.10 Max Fürbinger's (*A* and *B*) two views of "Experimental Family Trees of Birds," a three-dimensional phylogenetic tree of bird evolution, and "Horizontal (Planimetric) Projections of the Family Trees of Birds" through the (*C*) lower, (*D*) middle, and (*E*) upper portions of the tree, from *Untersuchungen zur Morphologie und Systematik der Vögel* (1888).

CHAPTER SIX

The Waning and Waxing
of Darwinian Trees

As the nineteenth century drew to a close, the reality of evolution became firmly ensconced within the scientific community. Darwin deserves credit for the over-used idea of a paradigm shift—evolution was a scientific "fact," but not so for Darwin's theory of natural selection. This intellectual retreat happened for a number of reasons. A common perception suggests that when Gregor Mendel's (1822–1884) work "Versuche über Pflanzen-Hybriden" (Experiments in Plant Hybridization, 1866) on particulate inheritance was rediscovered at the turn of the twentieth century, its sole importance came as a mechanism for inheritance. In the hands of scientists such as Hugo de Vries (1848–1935), ideas of genetic mutations provided not only the source of new genetic material but also the cause of proposed rapid evolutionary change, which for some biologists largely replaced Darwin's natural selection. This perception, while in part true, does not provide a complete answer.

Retreat from Darwin's natural selection began even before the rediscovery of Mendel and the birth of modern genetics. For example, one of the coauthors of natural selection, Alfred Russel Wallace, increasingly viewed some aspects of human faculties as mostly independent of natural selection, an exclusion likely caused in part by his later misguided acceptance of spiritualism. More important, Darwin could never provide an adequate hypothesis of inheritance or the source of variation on which his natural selection was to act. Darwin's self-described "much-abused" theory of Pangenesis was roundly dismissed.

Cope's Neo-Lamarckian Trees

These shifts of fortune had some effect on how evolution was represented in trees, but it was not because of work in genetics; rather, it came from the still emerging field of vertebrate paleontology, notably in the United States, and the ideas of evolutionary mechanisms that accompanied them. In 1868, only nine years after the publication of *On the Origin of Species*, an American vertebrate paleontologist was advocating a different mechanism for evolution in his paper "On the Origin of Genera." The author, Edward Drinker Cope (1840–1897), became well known outside of science because of one of the nastiest, most protracted confrontations in all of the history of science, the so-called bone wars that raged between Cope at the Academy of Natural Sciences in Philadelphia and Othniel Charles Marsh (1831–1899) at the Peabody Museum of Natural History at Yale University in New Haven. To make matters worse, Marsh was by and large a supporter of Darwinism, whereas Cope was not.

In his paper, the twenty-eight-year-old Cope (1868), another young Turk but never in Darwin's sphere, writes:

> That a descent, with modifications, has progressed from the beginning of the creation, is exceedingly probable. The best enumerations of facts and arguments in its favor are those of Darwin, as given in his various important works, The Origin of Species, etc. There are, however, some views respecting the laws of development on which he does not dwell, and which it is proposed here to point out. In the first place, it is an undoubted fact that the origin of genera is a more distinct subject from the origin of species than has been supposed. (243)

This provides a prime example of the acceptance of evolution, but not Darwin's version. Cope argued that genera, families, and other groups arose more through the retardation and acceleration in the process of development of an individual and that natural selection was subservient to this process. Whereas Cope's intuition about developmental timing would fit in today's understanding of its importance in the evolutionary process, his views on how this would affect higher categorical levels—genera, families, and so on—were completely misguided. Although not universal, the consensus is that evolution operates at and below the species. In and of itself, Cope's thesis was not Lamarckian, but he argued that the environment acted on the individual to cause these developmental changes, which were then passed to the next generation. Much of Cope's argumentation came from what he as well as colleagues, notably in the United States, believed they saw in the fossil record of vertebrates. This was termed neo-Lamarckism in 1884 by a contemporary and sometime colleague of Cope, Alpheus Hyatt (1838–1902) (Regal 2002 and references therein). The record for Cope, Hyatt, and others showed an orderly, almost rectilinear pattern of progression that did not accord well with Darwin's natural selection.

Cope produced nearly 1,500 scientific publications in his relatively short life, some a few lines long and others massive tomes, the most impressive of which is 1,009 pages of text plus almost as many additional pages, or at least an equal thickness of plates, many of them foldouts, published in 1884. This is a single, hardbound 12- by 9-inch (30 by 24 cm) quarto volume measuring an astounding 4¾ inches (12 cm) thick; no wonder it has earned the epithet "Cope's Bible." Many of the illustrations of specimens produced for Cope's publications were quite detailed and elaborate, such as in this ponderous 1884 tome, but the same cannot be said of his phylogenies, which were as simple as Haeckel's were grandiose. In this large volume, Cope provides only six tree-like diagrams, one for turtles and the remainder for mammals. The diagrams do not show consistency in orientation; some open upward, whereas others open downward. He identifies them by various names—tables, phylogenies, lines of descent. Their appearances, however, clearly conform to a general pattern, as seen by the three shown in figure 6.1*A–C*. The trees use Cope's terminology, and although there are many incorrect systematic aspects to them, they are not of concern here. Of greater interest are the regular spacing of names and the relative position of the names based on his perception of their grade of evolution. For example, in the downwardly opening tree (see figure 6.1*A*), among living members of his "Taxeopoda," the Perissodactyla (odd-toed ungulates) and Artiodactyla (even-toed ungulates) are placed farther along or more terminally, implying more evolutionary advancement compared with Proboscidea (elephants) and Hyracoidea (hyraxes). Similarly, the upwardly opening tree (see figure 6.1*B*) shows Equidae (horses) farther advanced than Rhinoceratidae (rhinos), which in turn occur farther along than Tapiridae (tapirs). Figure 6.1*C* shows successive steps in the lines leading to Felidae (cats) and Canidae (dogs). The repeated names of these two families indicate advanced living members along with more primitive extinct members of each family.

In his book *The Primary Factors of Organic Evolution* (1896), published the year before his death, Cope presents seven simple trees, much like those just discussed, but two of them differ in that they include geologic time. One offers a schematic phylogeny of plants (see figure 6.1*E*), and the second gives his views on mammalian evolution (see figure 6.1*F*). In the plant phylogeny, groups such as Cope's Protophyta (an obsolete term for a mix of groups sometimes including, for example, bacteria and blue-green algae) appear in the Ordovician period with successive appearances of eight other major plant groups, possibly coming from a common ancestor but then paralleling one another during their evolutionary and geological history. He argues in the pages preceding the plant phylogeny that this parallel evolution in various plant groups demonstrates what he terms "successional development," with changes in reproductive mode being particularly important. He sees two components for this, a trend going "from the simple to the complex" and a trend leading "from the generalized to the specialized." Cope used quotation marks for the three phrases, as shown here. He quickly notes that these concepts are not identical. Further, in the latter trend, he notes that both progressive and retrogressive processes occur, which relates to his earlier noted ideas of the importance of embryological development in evolution. For the mammalian phylogeny,

FIGURE 6.1 Edward Cope's evolutionary trees for (A) "Taxeopoda," including Perissodactyla (odd-toed ungulates), Artiodactyla (even-toed ungulates), Proboscidea (elephants), and Hyracoidea (hyraxes); (B) Equidae (horses), Rhinoceratidae (rhinos), and Tapiridae (tapirs); and (C) successive steps in the lines leading to Felidae (cats) and Canidae (dogs), from *The Vertebrata of the Tertiary Formation of the West* (1884). Cope summarized his ideas on development and evolution in (D) a very simple diagram, from "The Method of Creation of Organic Forms" (1871), and included geologic time in his schematic phylogenies of (E) plants and (F) mammals, from *The Primary Factors of Organic Evolution* (1896).

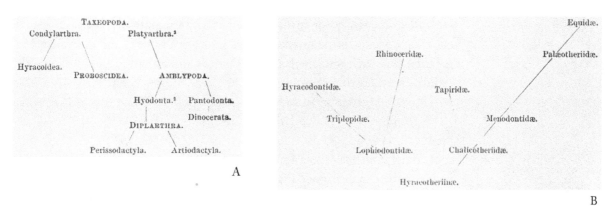

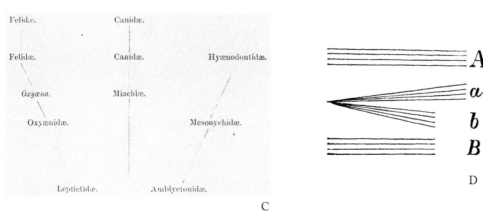

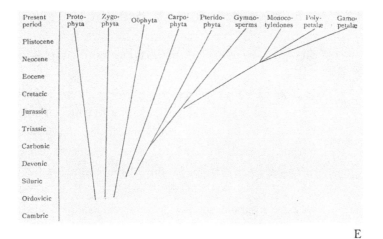

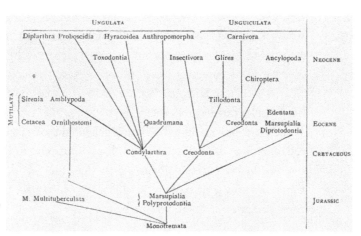

separate from greatly out-of-date systematic interpretations, his placement of various extant groups of mammals such as Chiroptera (bats) and Glires (rabbits and rodents) well back into the geological past is somewhat odd, until the pages preceding and following the phylogeny are read. Although not explicitly within the text, he clearly attempts to indicate when he believes these groups first appeared.

Examining Cope's theorizing on evolution provides some basis as to why he produced these as well as other trees as he did. In his 1868 paper, Cope did have some diagrams, but except for possibly one, none comes close to evoking a tree. In his slightly later paper "The Method of Creation of Organic Forms" (1871), he encapsulates views on development and evolution in a simple diagram at the end of the text (see figure 6.1D) that was repeated twice in his book-length treatment of his theory of evolution, *The Origin of the Fittest* (1887).

In the paragraph before the description of this small figure, Cope (1871) praises Darwin for demonstrating

> the origin of varieties in animals and plants, either in the domesticated or wild states . . . [and that] species have been derived from other species among domesticated animals . . . [and that inferences] that other species, whose origin has not been observed, have also descended from common parents . . . [is] to be justified; but when from this basis evolution of divisions defined by important structural characters, as genera, orders, classes, etc., is inferred, I believe that we do not know enough of the uniformity of nature's processes in the premises to enable us to regard this kind of proof as conclusive. (231)

He then goes on to explain the diagram on the same page:

> In *A* we have four species whose growth attains a given point, a certain number of stages having been passed prior to its termination or maturity. In *B* we have another series of four (the number a matter of no importance), which, during the period of growth, cannot be distinguished by any common, i.e., generic character, from the individuals of group *A*, but whose growth has only attained to a point short of that reached by those of group *A* at maturity. Here we have a parallelism, but no true evidence of descent. But if we now find a set of individuals belonging to one species, or still better, the individuals of a single brood, and therefore held to have had a common origin or parentage, which present differences among themselves of the character in question, we have gained a point. We know in this case that the individuals, *a*, have attained to the completeness of character presented by group *A*, while others, *b*, of the same parentage have only attained to the structure of those of group *B*. It is perfectly obvious that the individuals of the first part of the family have grown further, and, therefore, in one sense faster, than those of group *b*. If the parents were like the individuals of the more completely grown, then the offspring which did not attain that completeness may be said to have been *retarded* in their development. If, on the other hand, the parents were like those less fully grown, then the offspring which have added something, have been *accelerated* in their development.

Osborn's Aristogenetic Trees and the Rise of Neocreationism

Cope possessed a grasp of the importance of development in evolution, but his neo-Lamarckian views on how variation in development might arise never found traction among other biologists and paleontologists, with a few notable exceptions—in particular, one aspiring young vertebrate paleontologist, Henry Fairfield Osborn (1857–1935), who rose to considerable prominence in early-twentieth-century American science. Osborn would begin with Cope's neo-Lamarckian ideas but carry them even further and in the process produce some of the most elaborate tree-like diagrams of the early twentieth century.

Osborn studied anatomy at a New York hospital and embryology with Thomas Henry Huxley in England and received his doctorate in paleontology from Princeton. He also received his undergraduate degree at Princeton under the tutelage of Cope; there, he became fast friends with William Berryman Scott (1858–1947) and Francis Speir Jr. (1856–1925). With newly minted undergraduate degrees, the three undertook an expedition to fossil beds in the American West in 1877, which was written up and published (Osborn, Scott, and Speir 1878). Scott, like Osborn, went on to become a vertebrate paleontologist, whereas Speir earned a law degree in New York; but all three remained friends, and certainly Osborn and Scott early on influenced each other in their respective careers (Regal 2002). Scott even dedicated a later edition of *A History of Land Mammals in the Western Hemisphere*, first published in 1913, "to the memory of his friends" Speir and Osborn; Speir named one of his children Henry Fairfield Osborn Speir (Chamberlain 1900).

Cope clearly influenced Osborn and Scott. This can be seen in some of the phylogenies in their earlier works. In 1899, Scott published a paper, "The Selenodont Artiodactyls of the Unita Eocene," on the even-toed ungulates such as cows and deer that have crescent moon–shaped ridges (selens) on their teeth, with which they grind plant material much as tools used to file wood or metal. He provides a number of figures of the fossils and near the end one very simple but understandable phylogeny of the mammals discussed in his paper (figure 6.2*A*). Scott notes that this "subjoined table exhibits the relationships of the various genera as conjectured in the preceding pages." Other than the names being placed vertically, the simplicity in form matches well the trees of Cope (see figure 6.1). Even though similar to Cope's trees, Scott's style appears clearer in intent, not requiring explanation for clarity.

Fourteen years later, Scott (1913) presents a figure he titles "Evolution of the Proboscidea" (see figure 6.2*B*). He indicates modification after an earlier figure (Lull 1908) (see figure 6.2*C*). Scott's version adds lower molars on the left and outlined heads on the right. These are earlier renditions of this sort of representation. Scott shows a progression in the evolution of elephant heads and teeth. Although it is not strictly a *scala naturae*, it certainly approaches one. The tree shown in figure 6.2*D* comes from Osborn's published version of "The Rise of the Mammalia in North America" (1893), an address he presented to the American Association for the Advancement of Science the same year. He calls this very broad-brush phylogeny a "Hypothetical Phylogeny of the North American Mammalia." It is placed within the context of geologic time. Osborn makes some attempt to show the quality of the record and taxonomic richness by the thickness of the black columns.

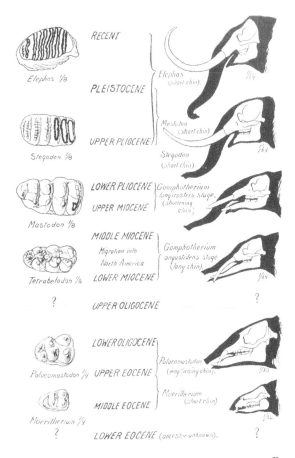

A

B

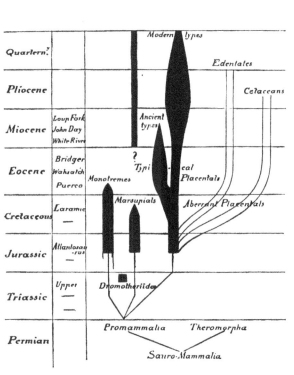

C

D

Recall that Edward Hitchcock (1840) and, even more similarly, Louis Agassiz (1844) had attempted a similar theme of using thicknesses of branches, albeit in a creationist context, to show relative taxonomic richness (see figures 3.10 and 3.11).

Later Osborn used other designs, such as phylogenies radiating within concentric circles, that echo in later figures. Based on his developing ideas of evolution, he often incorporated the idea of a concentric radiation with a series of radiating lines showing small outline figures of the animals in question. The popularity of this sort of representation has waxed and waned throughout the twentieth and early twenty-first centuries, notably in popular literature, where the public always remains fascinated with knowing what the animals looked like. A tree resplendent with many small vignettes of animals, especially those of long-extinct species, still holds great interest for the public. Two classic examples of Osborn's (1936) trees come from the first volume of his two-volume monograph on elephants and their relatives. Figure 6.3*A* is the "Phylogeny of Moeritherioidea, Deinotherioidea, and Mastodontoidea," all of which are extinct relatives of elephants, and the flame-like tree in figure 6.3*B* is titled "Final Diagram (1935) Showing the Adaptive Radiation of the Forty-three Generic Phyla of the Proboscidea." Although the phylogeny in figure 6.3*B* does not include small outlines of animals, figure 6.3*A* provides classic Osborn elements: it radiates from an ancestral point with a series of separate lineages, and each has a vignette outline of included taxa.

Recall that Osborn was influenced by Cope's ideas of neo-Lamarckian evolution, and he supplemented Darwin's idea of natural selection with what he called aristogenesis, a *scala naturae*–like, purpose-driven concept of evolution that can be seen in many of his phylogenies, such as the two in figure 6.3*A* and *B*. Such figures are often dominated by straight-line evolutionary change (see also figure 1.7). No matter what might be the stated intent, Osborn's views, as seen in many of his phylogenies, show much of evolutionary modification occurring along a single line of change and descent, hence the clear resemblance to the older *scala naturae*, or great chain of being. At the time, and for a number of years to come, his ideas had a considerable impact because from 1908 until 1933 Osborn directed the American Museum of Natural History, still one of the premier natural history museums in the world, but at that time an even more powerful institution in American science, in large measure because of Osborn. Osborn's wealth and connections helped amass power and prestige for his museum, ensuring his place in the history of that institution, at the time a very influential purveyor of scientific ideas. Today, his reputation as an excellent administrator remains but not so his scientific reputation,

FIGURE 6.2 William Scott's (*A*) stratigraphically arranged phylogeny, from "The Selenodont Artiodactyls of the Unita Eocene" (1899), and (*B*) "Evolution of the Proboscidea," including lower molars to the left of the skulls and outlined heads to the right, from *A History of Land Mammals in the Western Hemisphere* (1913), based on (*C*) a similar figure by Richard Lull, from "The Evolution of the Elephant" (1908); (*D*) Henry Osborn's "Hypothetical Phylogeny of the North American Mammalia," placed within the context of geologic time, with some attempt to show the quality of the record and taxonomic richness by the thickness of the black columns, from "The Rise of the Mammalia in North America" (1893).

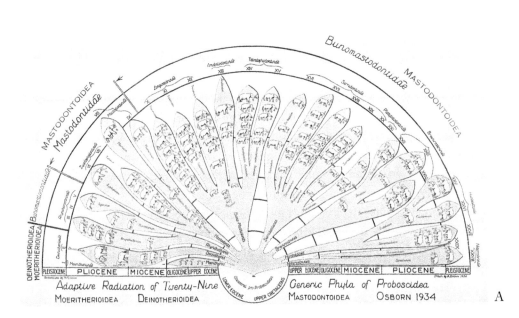

Adaptive Radiation of Twenty-Nine

MOERITHERIOIDEA   DEINOTHERIOIDEA

Generic Phyla of Proboscidea

MASTODONTOIDEA   OSBORN 1934

A

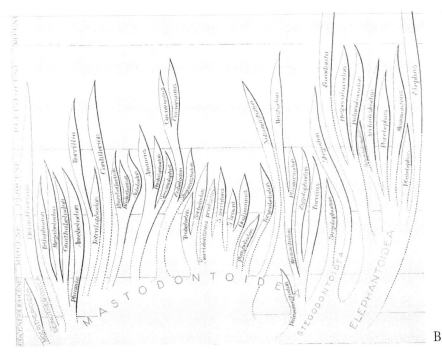

B

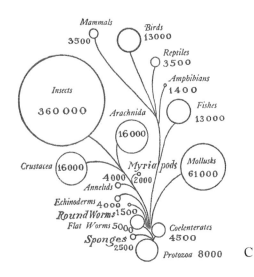

C

in part because of his support of teleologically tainted neo-Lamarckian ideas of evolution and in part because of his racially driven ideas of eugenics (Rainger 1991; Regal 2002).

Osborn and, before him, Cope were two of the most dominant scientists in influencing tree building in the late nineteenth century and the first quarter of the twentieth, but it was a trial including an evolutionary tree that truly captured the attention of the American public, later spawning a play, a film nominated for four Academy Awards, and a variety of books. The trial is best described by Edward Larson in *Summer for the Gods* (1997), which details the machinations leading up to and through the "Scopes Monkey Trial," which took place in Dayton, Tennessee, in 1925. The Tennessee state legislature had passed the Butler Act, which made it "unlawful for any teacher . . . to teach any theory that denies the Story of the Divine Creation of man as taught in the Bible, and to teach instead that man has descended from a lower order of animals" (Tennessee House Bill 185, 1925). As Larson relates it, the high-school teacher John T. Scopes reluctantly agreed to test the law and in so doing helped put Dayton on the map, as well as helping a fledgling American Civil Liberties Union defend one of its first cases.

A state-approved textbook, George William Hunter's *A Civic Biology* (1914), became exhibit number one in the case (the Bible was number two), although it really served only as a straw man for the case. It is also mentioned in Jerome Lawrence and Robert Edwin Lee's play *Inherit the Wind* (1955), based quite closely on the trial transcript, as well as in Scopes's memoir (Scopes and Presley 1967). In a number of places, Hunter unquestionably praised and supported Darwin's evolutionary ideas, but he also espoused racist and eugenic views. On page 193 of Hunter's text is an orthogenetic-like diagram of horse evolution after one by the paleontologist William Diller Matthew (incorrectly called Matthews), and on page 194 is what Hunter terms an evolutionary tree (see figure 6.3C). Each branch of this quite generalized, *scala naturae*–like tree terminates in a circle, the circles varying in size according to the relative number of species they include, which is given within or next to each circle. The text speaks only obliquely of human evolution, stating that man at first was "little better than one of the lower animals" (196), but the tree presents nothing about human evolution or human descent from lower forms. This did not stop prosecutors in the trial from attempting to portray the book and, by implication, the evolutionary tree as showing human descent from lower forms, so that Scopes's use of the text in class could be used to find him guilty of having violated the Butler Act. The trial transcript (1990:123) specifically refers to the evolutionary tree in a question by the prosecution regarding whether Scopes discussed it in his class.

FIGURE 6.3 Osborn's (A) "Phylogeny of Moeritherioidea, Deinotherioidea, and Mastodontoidea," all of which are extinct relatives of elephants, showing radiation from an ancestral point with a series of separate lineages, each with a little vignette outline of included taxa, and (B) "Final Diagram (1935) Showing the Adaptive Radiation of the Forty-three Generic Phyla of the Proboscidea," from *Proboscidea* (1936); (C) the phylogeny in George Hunter's *A Civic Biology* (1914) placed into evidence in the "Scopes Monkey Trial" (1925), as well as being mentioned in the play *Inherit the Wind* (1955) and the film of the same name (1960), based on the trial. ([A] and [B] reproduced with permission of the American Museum of Natural History)

Probably wanting to deflect any further scandal for his textbook, in his revised edition, which appeared only one year after the trial, Hunter (1926) expunged not only the five-page section on evolution but also any mention of the word, as well as the horse evolution chart and the evolutionary tree. This recalls Hitchcock's (Hitchcock and Hitchcock 1860) deletion of his nonevolutionary, tree-like pale-ontological chart (see figure 3.11) when the tree became an evolutionary icon following the publication of *On the Origin of Species* in 1859. The back-peddling and kowtowing to the religious right signaled a change that gained momentum in the middle of the twentieth century. The trial seemed to damp creationism for a few decades, but it returned with a vengeance in the mid-twentieth century, especially in the Unites States, along with its later spawn "intelligent design." This time, unlike Hitchcock and Agassiz in the nineteenth century, who honestly attempted to wedge biblical creation into a scientific mold, the newcomers promoted a pseudoscience that still festers like an intellectual canker.

The New Synthesis and Neo-Darwinian Trees

Cope and Osborn influenced how the next academic generations represented biological order, but the same cannot be said of their neo-Lamarckian evolutionary ideas, which soon faded. Many trees were produced in the first two-thirds of the twentieth century, so the task of choosing among them becomes daunting, but three academic descendants and associates of Osborn in particular typify what came next.

William King Gregory (1876–1970) obtained his doctorate from Columbia University and assisted Osborn in his research and in the preparation of his publications. Gregory concentrated on the evolution and comparative anatomy of vertebrates, especially mammals. Two other persons of note with ties to the American Museum of Natural History who effectively used phylogenetic trees came in the academic generation following Gregory: Alfred Sherwood Romer (1894–1973) and George Gaylord Simpson (1902–1984). Romer was a student of Gregory's, earning his doctorate at Columbia and then spending much of his career at Harvard, whereas Simpson, after obtaining his doctorate from Yale, spent his career in part at the American Museum of Natural Museum and at Harvard. Romer and, more so, Gregory produced copious phylogenetic trees. Romer is best known for those in his text *Vertebrate Paleontology*, which went through three editions and a number of printings between 1933 and 1971; Gregory compiled many such figures in *Evolution Emerging* (1951). Simpson did not produce nearly as many phylogenetic trees of specific groups as did Gregory and Romer, but he was one of the first and still most effective users of phylogenetic trees in the explanation of evolutionary and ecological theory and process.

Examining the contributions of Gregory, Romer, and Simpson in visualizing biological order, we start with the senior member of the group, William King Gregory. Gregory's career began as an assistant to Osborn near the turn of the twentieth century. Osborn worked all his assistants very hard, but none more than Gregory. Osborn's two best-known works—a two-volume study of titanotheres, distant relations of rhinos and horses (1929), and the even better-known two-volume monograph on proboscideans (1936, 1942)—can in large measure be traced to the work of Gregory. Gregory also contributed greatly to the writing of Osborn's *Age of Mammals in Europe, Asia and North America* (1910; Colbert 1975). If Gregory ever supported the Cope–Osborn views of evolution, by the time he produced his magnum opus in 1951, *Evolution Emerging*, based on a half century of work, he largely supported the newer views set forth in the Modern Synthesis, which unified Darwinian natural selection and Mendelian genetics: "Perhaps the greater part of the causes of evolution lies deep within the nature of the hereditary mechanism and of its reactions to Natural and Artificial Selection; but . . . the leading principles of evolution . . . appear to operate all levels of organization, from inorganic material . . . along the diverging paths of descent with modifications" (1:552). Simpson called these two volumes "both the chef d'oeuvre and the swan song of a genius" (quoted in Colbert 1975).

Out Haeckeling Haeckel

Hands down, Gregory was the most prolific purveyor of phylogenetic trees, even bettering the number produced by Ernst Haeckel (Pietsch 2012). The first volume of *Evolution Emerging* (1951) presents 736 pages of text on all animals but emphasizing vertebrates; while important and worth the read, it is secondary

to the second volume, whose 1,013 pages feature wondrous illustrations of all manner of animals. Except for a very imaginative, and dare I say Haeckelian, geologically based tree frontispiece titled "Procession of the Vertebrates" (figure 6.4), the first volume lacks illustrations. No images of animals are shown in the diagram, but the spindle-like aspect gives the sense of waxing and waning of various lineages, harking back to Agassiz's creationist diagram of fish relationships produced over one hundred years earlier (see figure 3.10). Instead of the tree being anchored by a geologic time scale on one side, the vertebrate procession emerges from the depths of a Grand Canyon–like chasm; thus the appropriate name of the book, *Evolution Emerging*.

The second volume includes almost ninety trees of various forms. In addition, Gregory presents diagrams showing a succession of species or parts of their anatomies in an almost *scala naturae* fashion. We can only sample a few to provide a sense of what he attempted to show. Even though Gregory was not a fan of Osborn's views on evolution, the influence of Osborn comes through in a number of Gregory's trees. Excluding the trees that he presents from other sources, a majority of his trees show whole organisms (extinct and extant), skulls of these organisms, or other parts of the organisms. In most of the figures, the illustration of the animal or its parts is large enough to see the anatomical

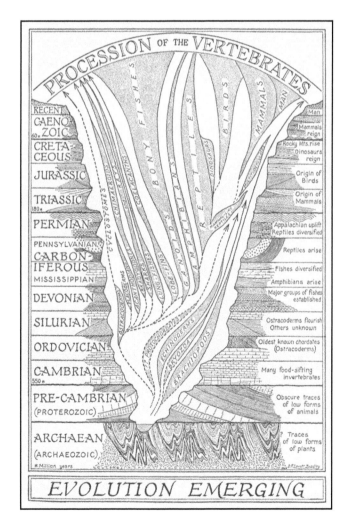

aspects that Gregory wished to show his readers. The actual tree portion is often secondary, but he nonetheless was attempting to depict what he called, and we still call, adaptive evolution or radiation. This came from Gregory's insistence on the importance of understanding the integration of the various parts of an animal's body. Above all, Gregory was one of the twentieth century's finest comparative anatomists. He understood what many others did not—that parts of an organism do not evolve in lock step. Rather, different parts evolve in fits and starts; thus the creationist cry of how could a bird fly with only half a wing becomes patently absurd. This idea of different rates of evolutionary change in different parts of an animal's body is now known as mosaic evolution, with the obvious allusion to the parts of a mosaic creating the whole. Gregory recognized this, calling it his "palimpsest theory" in which later specializations overlie earlier ones, just as the term is used in the original meaning of the overwriting of older by younger texts. With this in mind, Gregory's considerable interest in presenting phylogenies of only parts of an organism becomes obvious. Many examples occur in the second volume of *Evolution Emerging*, such as the evolution of marsupials in general using whole-body outlines (figure 6.5*A*) and, in particular, the evolution of the hind feet in Australian marsupials (see figure 6.5*B*) and of the occlusal, or biting, surface of a typical lower (see figure 6.5*C*) and upper (see figure 6.5*D*) molar. Gregory is attempting to show that different parts of an organism evolve at different rates and must be considered separately, but then he puts them back together as a whole to comprehend the adaptive radiation of that group.

Gregory also often employs a different sort of diagram. At first glance, it appears to be some form of *scala naturae*, but it is not. Figure 6.6*A* shows a group of skulls suggesting the march from primitive placental mammal forms to advanced elephants; they are very similar to figure 6.2*B* and *C*, which Scott (1913) and Richard Lull (1908), respectively, called evolutionary diagrams of elephants. Instead, Gregory calls this a "structural series, showing recessions of nasals"— meaning retraction of the nasal opening backward and upward onto the skull. Note the vertical line demarcating the front and back halves of the skull. Unlike earlier authors such as Lull, Scott, and Osborn, Gregory does not imply or state an evolutionary succession but rather the likely structural changes that occurred as elephants evolved. Another example shows the "general relations of the temporal and masseter muscles . . . from fish to man" (see figure 6.6*B*). The arrows track from the shark, then over, and finally down to humans. Once again, although Gregory would no doubt regard a shark as primitive relative to a human, the intent in his pseudo-ladder shows changes in the muscles of the jaw as evolutionary history unfolded, much as in his palimpsest style, but with no implication of an evolutionary series.

The Master of the Textbook Tree

Next follows Alfred Sherwood Romer, a student of Gregory's at Columbia near the end of the second decade of the twentieth century. Romer learned comparative vertebrate anatomy from the master, which obviously influenced his work

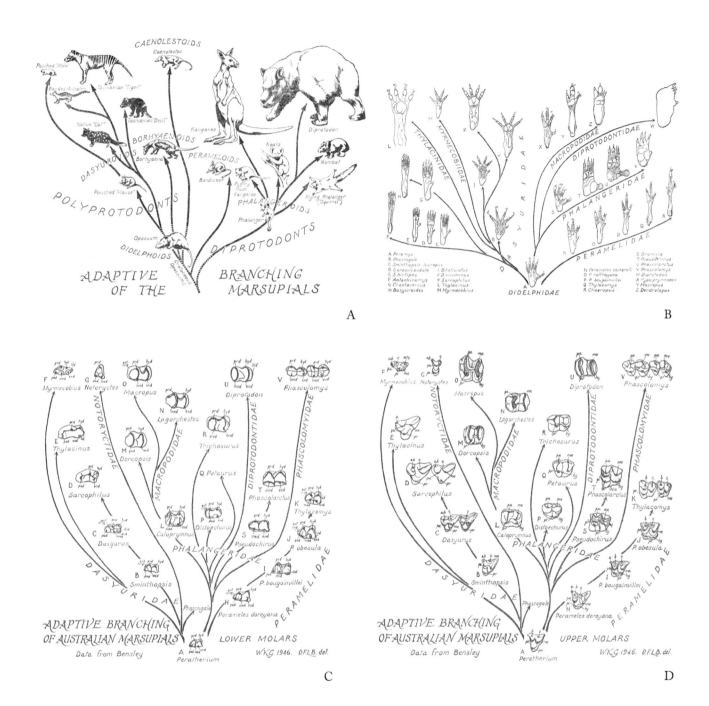

FIGURE 6.5 Gregory's four phylogenies of marsupials, showing (A) the evolution of marsupials in general using whole-body outlines, (B) the evolution of the hind feet of Australian marsupials in particular, and the occlusal, or biting, surface of a typical (C) lower and (D) upper molar, from *Evolution Emerging* (1951). (Reproduced with permission of the American Museum of Natural History)

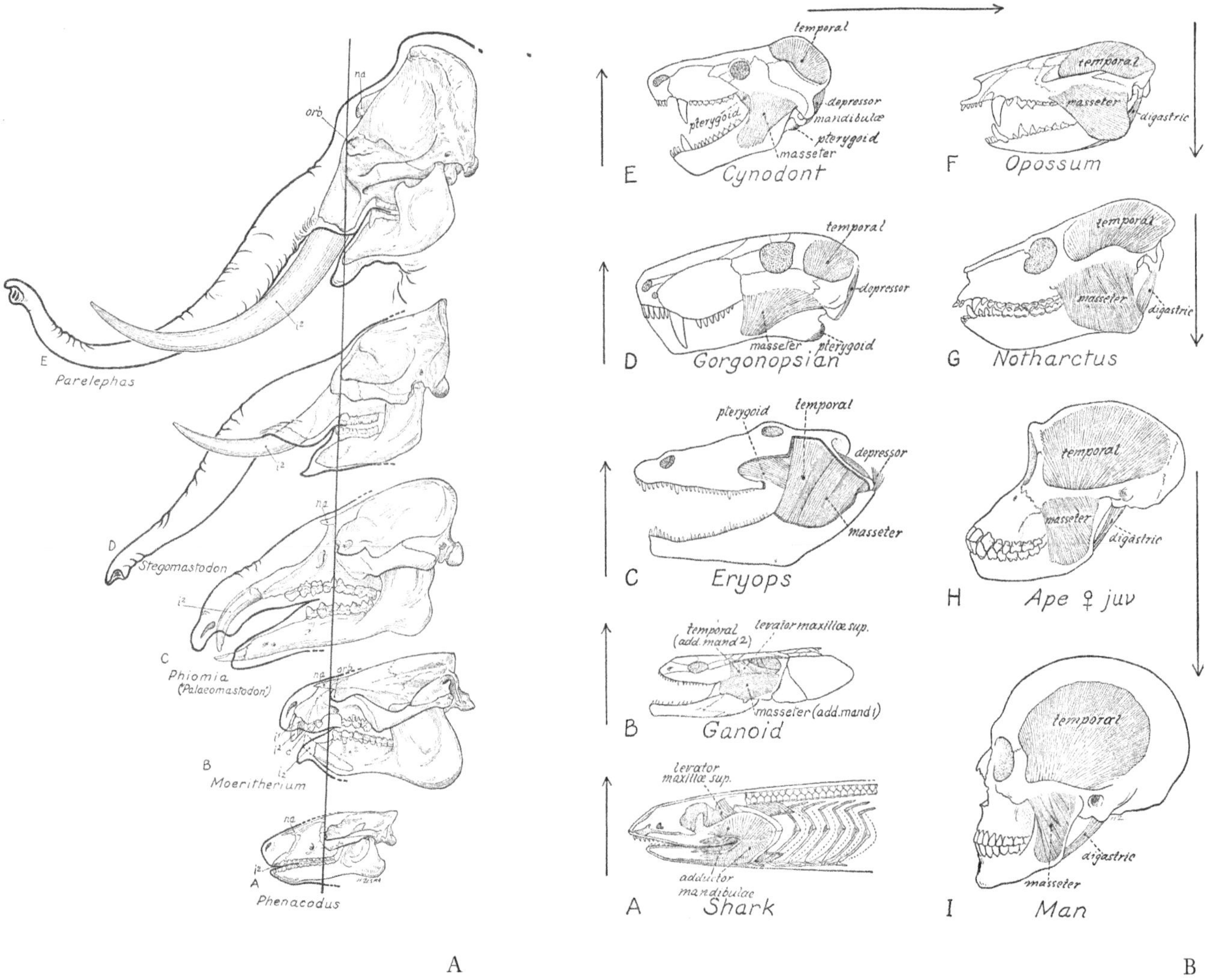

A

B

throughout his career in the study of early tetrapod evolution, including the lineages that would lead to reptiles and the others that eventually led to mammals (Colbert 1982). As with his mentor, Romer was a prolific producer of phylogenies. Among his numerous papers and books, the most popular were his textbooks *The Vertebrate Body* (later editions done with Thomas Parsons) and *Vertebrate Paleontology*, used by many students, myself included. Both have about fifteen trees of various sorts; most in *The Vertebrate Body* consist of lightly sketched tree outlines on which Romer placed small drawings of the animals, with no geologic time shown, whereas those in *Vertebrate Paleontology* consist of spindle diagrams with no animals shown but within the context of geologic time.

Figure 6.7*A*, from the fifth edition of *The Vertebrate Body* (1977), shows the "Diagrammatic Family Tree" of eutherian (placental) mammals, with an animal sketch representing each lineage. Figure 6.7*B* and *C* show the "Chronologic Distribution of Placental Mammals" from, respectively, the first edition (1933) and the third edition (1971 printing) of *Vertebrate Paleontology*. The first tree shows almost

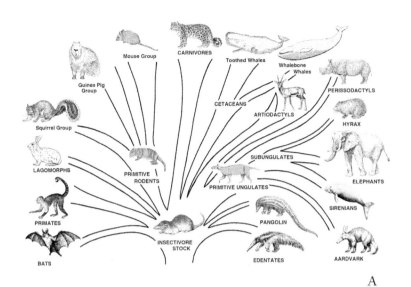

A

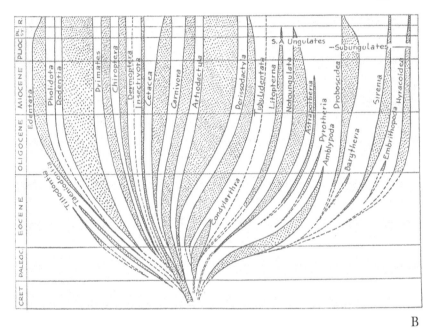

B

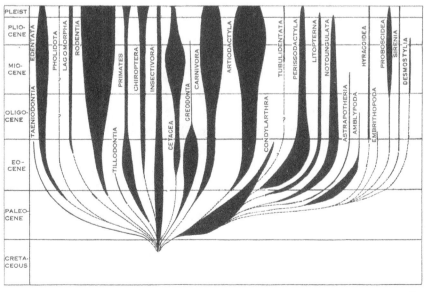

C

FIGURE 6.7 (A) Alfred Romer and Thomas Parsons's "Diagrammatic Family Tree" of placental mammals, redrawn from *The Vertebrate Body* (1977); Romer's (B) "Chronologic Distribution of Placental Mammals," from *Vertebrate Paleontology* (1933), and (C) "Chronologic Distribution of Placental Mammals," from *Vertebrate Paleontology* (1971). ([B] and [C] reproduced from Romer 1933, 1971; used with permission of the University of Chicago Press)

no extinct forms, whereas in the other two trees a number of extinct lineages appear. Although the latter two do vary in some of the relationships represented, they otherwise look rather similar except for differences in shading. The distinction between the first tree and the other two trees relates to the intended audience. For *The Vertebrate Body*, the audience was students of comparative anatomy interested in living forms, whereas for *Vertebrate Paleontology*, the audience comprised students of that discipline and hence their interest in more extinct forms.

Other trees showing comparable groups in the 1933 and 1971 editions differ markedly, notably those for rodents. In 1933, only ten superfamily or family-level groups appear (as well as rabbits, the duplicidentates) (figure 6.8*A*), compared with forty-four similarly named lineages in 1971 (see figure 6.8*B*). This change came from increased knowledge of rodents as well as from Romer's desire simply to show more detail. Although Romerian in style and influential for the work of others, none of these trees broke new ground in tree construction, yet one of the smallest and simplest trees, the same in all three editions of *Vertebrate Paleontology* beginning with the first edition in 1933, provides some idea of how Romer and others were grappling with the issues of relating tree construction with classification. In this tree, Romer attempts to clarify how the concepts and differences between "vertical" and "horizontal" classification appear in a phylogenetic tree, portending the revolution in biological systematic thinking to soon come—phylogenetic systematics or cladistics (see figure 6.8*C*). In both 1933 and 1971, Romer writes:

> Two types of classification are possible—"vertical" and "horizontal." . . . Under the first system each family or other unit comprises all members of a known line from its first beginnings to its end or to modem times; the cleavage between lines is carried down to the very base of the evolutionary tree. But when, for example, forms are discovered seemingly ancestral to two distinct families or closely related to both, their inclusion in one or the other seems improper. Under such circumstances the best solution seems to be a "horizontal" cleavage, the erection of a stem group, including the base from which the long-lived later families may been derived. (1933:7, 1971:4)

In the 1971 edition, Romer included this additional paragraph:

> "Horizontal" or "vertical" classifications of the sort just described are both acceptable. To be avoided, however, is another type of "horizontal" classification. Let us assume that of the forms shown in the diagram, A and B, although with separate pedigrees, both evolved in each the same direction. If, after setting up (as at the right of the diagram) a basal family C, a scientist

FIGURE 6.8 Romer's evolutionary trees for rodents (and rabbits), going from (A) only ten superfamily or family-level groups in 1933 to (B) forty-four similarly named lineages in 1971, and (C) diagram clarifying how the concepts and differences between "vertical" and "horizontal" classification appear in a phylogenetic tree, from *Vertebrate Paleontology* (1933, 1971). (Reproduced with permission of the University of Chicago Press)

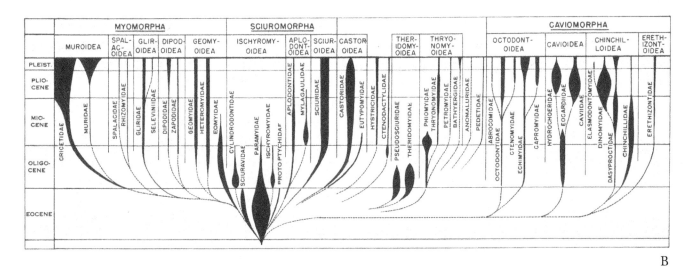

B

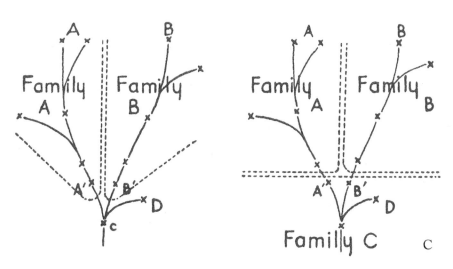

C

decides, because of similarities, to unite A and B in a common family, he has created a family that is not merely "horizontal" but polyphyletic—that is, he has included forms of different origin in a common assemblage. Sometimes, because of inadequate knowledge, this has been done, inadvertently, by pale-ontologists, but such unnatural grouping should be avoided. (4)

Biological classification provides perennial problems, but the issue became par-ticularly acute beginning in the mid-twentieth century. Romer voices this con-cern with an obvious expansion of these issues in the final edition of *Vertebrate Paleontology*.

A Synthesizer par Excellence

Recall that Darwin's single foldout diagram in *On the Origin of Species* (1859) did not attempt to explain a specific set of evolutionary relationships for a specific group of species but tried to express visually how he thought his theory of natural selection might explain the evolutionary process over great expanses of geologic time. In the first half of the twentieth century, it was George Gaylord Simpson more than anyone else who followed in Darwin's tradition of using phylogenetic trees to deal with evolutionary processes and theory, including issues of taxonomy such as those raised by Romer.

Along with a number of mostly British and American scientists, Simpson helped to forge the Modern Synthesis of evolutionary theory, also know somewhat inaccurately as neo-Darwinism. Scientists from the Darwinian camp (who were often natural historians) and Mendelian geneticists remained at loggerheads until population geneticists and population biologists in the 1930s and 1940s demon-strated through statistical studies of plant and animal populations that Darwin's natural selection winnowed the variation of Mendelian genetics. Darwin's natural selection proved instrumental in sorting variations, but Mendelian genetics pro-vided these variations that were altered by mutations. Simpson argued that the Modern Synthesis could be demonstrated through studies of the fossil record, and many tree-like figures centered on explaining this process. Simpson used tree-like diagrams to explain how the evolutionary processes unfolded, sometimes with hypothetical examples and at other times using case studies from the fossil and modern records. The examples explored here come from one scientific paper and his more popular books stretching through the middle third of the twentieth century.

In 1937, Simpson published an article ahead of its time, "Patterns of Phyletic Evolution," in which he examined several knotty issues especially in the light of the emerging Modern Synthesis of evolution. How does one interpret evolutionarily and taxonomically a "sample of similar but varying specimens extending over an appreciable span of geologic time" (1937a:307)? This hypothetical sample is given in figure 6.9*A*, along with Simpson's three interpretations gathered by his biographer Leo Laporte (2000). Simpson provides three scenarios: the first "representing two distinct and parallel phyla, with the type of frequency distribution at any one time, such as T, that would make this interpretation probable" (308) (see figure 6.9*B*); the

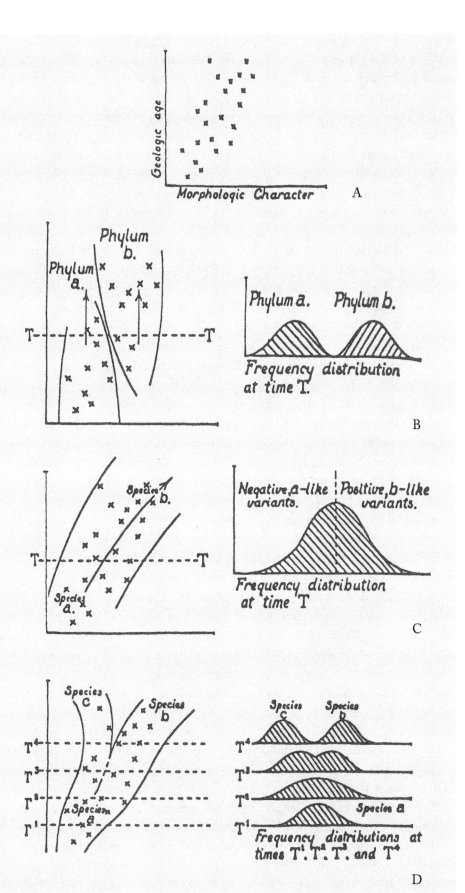

FIGURE 6.9 George Gaylord Simpson's (*A*) sample and (*B–D*) scenarios for the evolution of lineages over geologic time, from "Patterns of Phyletic Evolution" (1937). (Reproduced with permission of the Geological Society of America)

second "representing a single changing, with the type of frequency distribution at any one time, such as T, that would make this interpretation probable" (309) (see figure 6.9C); and the third "interpreted as an ancestral stock giving rise to two divergent phyla with the types of frequency distributions at successive times T^1, T^2, T^3, T^4, that would make this interpretation probable" (311) (see figure 6.9D). Of course, Simpson uses the word "phylum" here to indicate a general lineage or branch rather than the formal taxonomic rank of phylum. Later in his career, Simpson credited this paper with his abandonment of typological thinking for a more statistically biometrical approach to evolutionary theory and taxonomic issues (Laporte 2000). In the same year, Simpson followed with papers that examined in a statistical framework Paleocene and Eocene fossil mammals, most notably his Crazy Mountain mammal monograph (1937a), one of the earliest uses of biometrics looking at fossil populations. Typological thinking was giving way to populational thinking in evolutionary studies, but except in the hands of scientists such as Simpson, tree-like diagrams tended not to embrace such thinking.

Simpson's method of explaining complex evolutionary theory frequently employed simple diagrams, including tree-like figures (Laporte 2000). He was a master at the use of these almost chalkboard-like sketches. There are many from which to choose, but a few shown here exemplify some key issues, starting with arguably his most influential and at the time most controversial book, *Tempo and Mode in Evolution*, first published in 1944. In this volume, we see his visual musings on how he perceives evolution. Because he was one of the architects of the Modern Synthesis of evolution, which combined Darwin's natural selection with Mendel's genetics, his figures must be viewed through this prism.

In 1972, Niles Eldredge and Stephen Jay Gould proposed the theory of "punctuated equilibrium," which argues that most evolutionary change (usually meaning morphological change in fossils) occurs during rapid bursts of speciation followed by long periods of stasis when rates of evolutionary change and speciation are much lower (figure 6.10A). Simpson proposed a precursor to punctuated equilibrium in *Tempo and Mode*. Figure 6.10B shows his hypothetical phylogeny of this process. Simpson (1944) called this "explosive" (his quotation marks) evolution in which there are "multiple quantum steps into varied adaptive zones, followed by extinction of unstable intermediate types and phyletic evolution in each zone" (213). By phyletic evolution, Simpson meant gradual change mostly in a single lineage, or what we now call anagenesis. Punctuated equilibrium and "explosive" evolution models arguably differ in that the former addresses what might be called normal evolution, whereas the latter addresses the origin and rapid radiation of major groups in a short period of time. We do know more clearly now that evolution occurs at and below the species level, whether it results in so-called normal evolution or a major radiation over longer intervals of time. Possibly this is a distinction without a difference; nevertheless, Simpson's tree motif explores this concept.

A final set of Simpson's trees showing the adaptive radiation of primates tells an interesting story of changing ideas, specifically for Simpson and generally for evolutionary biology, regarding interpretations of evolution and the nature of taxonomy in the mid-twentieth century. These primate phylogenies come from Simpson's books *The Meaning of Evolution* (1949) (figure 6.11A) and *Principles of*

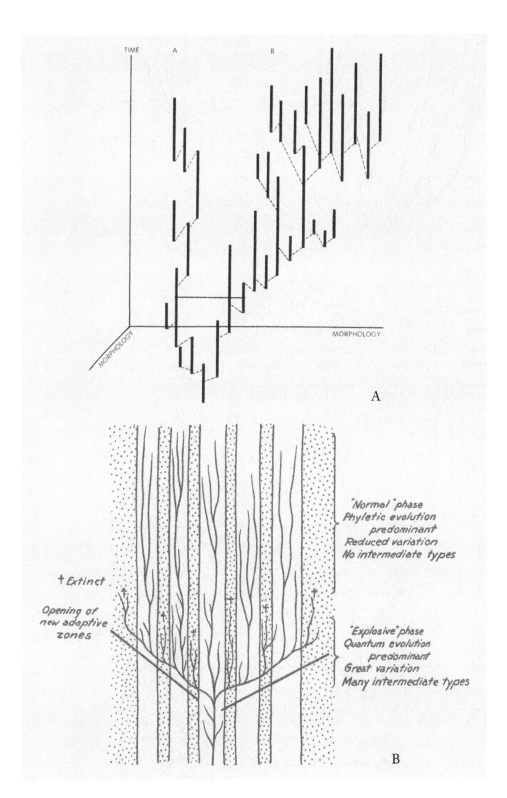

TIME

A   B

MORPHOLOGY

MORPHOLOGY

A

"Normal" phase
Phyletic evolution
predominant
Reduced variation
No intermediate types

† Extinct

Opening of
new adaptive
zones

"Explosive" phase
Quantum evolution
predominant
Great variation
Many intermediate types

B

FIGURE 6.10 (A) Niles Eldredge and Stephen Jay Gould's diagram of what they termed "punctuated equilibrium," from "Punctuated Equilibria" (1972); (B) Simpson's figure depicting a similar idea, which he called "explosive" evolution, modified from *Tempo and Mode in Evolution* (1944). ([A] reproduced with permission of Niles Eldredge)

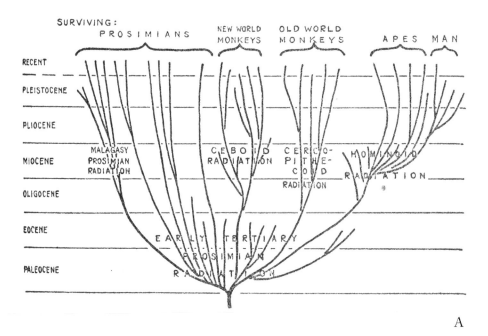

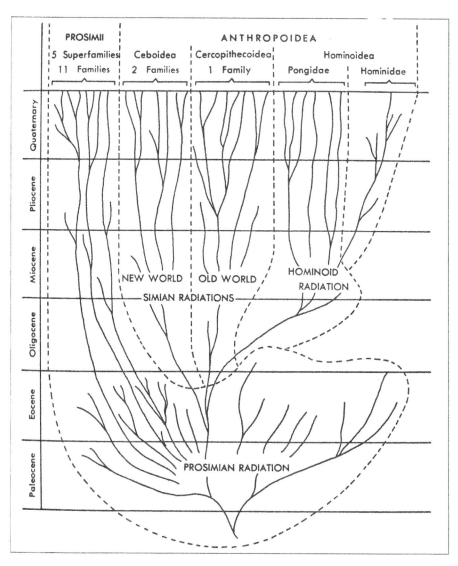

Animal Taxonomy (1961) (see figure 6.11*B*). These two trees, now outdated on several levels, show primate evolution as viewed at the time. The two phylogenies are similar in general pattern but differ in two important ways. First, the 1949 tree shows a noticeably staccato appearance of the four labeled groups, whereas the 1961 tree evens out these sharp breaks. The 1949 representation is closer to Simpson's view in *Tempo and Mode* (1944), in which evolution was regarded as happening both rapidly and slowly. By 1961, the revolt against the early-twentieth-century view of rapid, mutational change advocated by the early Mendelian geneticists was over. An interpretation—whether incorrect or not—of slow, stately evolutionary unfolding was in vogue. Simpson's (1961) phylogeny reflects this perspective, whether intended by Simpson or not. He had been criticized for advocating rapid evolutionary change in *Tempo and Mode*, and his later writing and his trees reflect this. The second difference between these two phylogenies comes in the form of the suggested taxonomies, echoing the vertical versus horizontal classifications discussed by Romer (1933, 1971). In the two phylogenies, prosimians (a now discarded taxon) are shown both as a grade of early primate and as a clade or lineage. The phylogeny from 1949 recognizes four major named radiations of primates. The phylogeny from 1961 shows this but also puts a dotted line around prosimians, making them both a grade (horizontal) and a clade (vertical) of primates.

The trees of the mid-twentieth century, with those of scientists such as Simpson leading the charge, emphasized the grand unfolding radiations of life over geologic time, very much in accord with perceptions of a Darwinian view of evolution. But perceptions and research programs in the 1970s, including how to build and show trees, were about to change radically for three reasons: first, the emergence of a new school of phylogenetic reconstruction; second, the development of vastly increased computer power and algorithms; and third, the explosion of molecularly based phylogenetic analyses primarily by dint of a procedure called polymerase chain reaction (PCR) that made the analysis of vast quantities of genetic information possible. Our perceptions of evolution, how we show this in trees, and even our view of ourselves within this scheme would never be the same.

FIGURE 6.11 Simpson's trees separated by a decade and reflecting changing ideas about the adaptive radiation of primates, going from (*A*) a noticeably staccato, bunched appearance of the four labeled groups, from *The Meaning of Evolution* (1949), to (*B*) an evening out of the sharper breaks, from *Principles of Animal Taxonomy* (1961). ([*A*] reproduced with permission of Yale University Press; [*B*] reproduced with permission of Columbia University Press)

CHAPTER SEVEN

Three Revolutions in Tree Building

The trees of the mid-twentieth century, especially those of scientists such as Alfred Sherwood Romer and George Gaylord Simpson, emphasized the grand unfolding radiations of life over geologic time, very much in accord with perceptions of a Darwinian view of evolution. By the 1970s, perceptions and research programs, including how to build and show trees, began to change radically for three reasons. First, in the 1960s two new, quite different schools of biological systematics emerged that challenged how one assesses relationships and how one represents these relationships visually. Second, although biochemically and genomically based phylogenetic studies began before the invention of the polymerase chain reaction (PCR) procedure in 1983, this invention was a watershed that made the analysis of vast quantities of genetic information possible. In addition, an apparent threat to tree iconography emerged as molecular studies expanded. Simple organisms possess fewer anatomical characters than more complex organisms that might be useful in systematics studies, and these simpler organisms also left less of a fossil record. More powerful molecular tools permitted a more thorough comparison of these simpler organisms, but they also threatened the very tree iconography by arguing instead that life's history represented an interwoven web or Darwinian tangled bank. Third, computing power increased by leaps and bounds in the latter half of the twentieth century, making the analysis of vast data sets possible. These three changes affected how we perceive evolution, how we show this on trees and other sorts of diagrams, and even how we ourselves, within this scheme, changed forever—all of which made for a true scientific and technical revolution.

Three Schools

No matter how one wishes to define it, much of the study of evolution is as a historical science. The origin of life on planet Earth happened once, or once at least as we suspect. At any rate, this was a singular event or maybe a series of events in which we have participated and which we have observed. This means that uncovering the process and pattern of what until at least recently entailed a paleontological and anatomical approach lay outside of what was regarded as scientifically testable. Even giants of evolutionary tree construction such as Ernst Haeckel, William King Gregory, Romer, and Simpson in the end used their superior knowledge of the subject to argue that this or that tree was likely correct. This science done by appeal to authority became known as "evolutionary systematics" compared with the newly arising and arguably more testable schools of "numerical taxonomy" and "phylogenetic systematics."

Beginning in the 1960s, this appeal to authority began to wane, and no less an authority than Simpson realized this shift in ideas. In *Principles of Animal Taxonomy* (1961), he writes, "I shall here acknowledge a debt to Hennig (1950), which is certainly one of the most valuable books on taxonomy . . . that has yet appeared," but he notes that his agreement with Hennig's approach may be "only partial, but it is substantial" (71n.2). He was referring to Willi Hennig's (1913–1976) then obscure and difficult book *Grundzüge einer Theorie der phylogenetischen Systematik* (*Outlines of a Theory of Phylogenetic Systematics*, 1950), about which the great evolutionist Ernst Mayr (1982) noted that it "is written in rather difficult German, some sentences being virtually unintelligible" (226), a rather damning statement from a German-born and German-educated American biologist. Mayr was not alone in his assessment of the book. Additionally, Hennig did not seem to know or at least did not cite much of the work of other great systematists of the time; nevertheless, Simpson must be given credit for his prescient take on the significance of Hennig's work. In his autobiography, Simpson (1978) expounded less kindly about Hennig's ideas, likely because by then the Hennigian school of systematics had continued its ascendency and with it various methods of visualizing evolutionary relationships. Much of this eventual success can be attributed to the translation, clarification, and rewriting of Hennig by Dwight Davis and Rainer Zangerl in *Phylogenetic Systematics* (1966), also the eponym of the Hennigian school. Even with Simpson's (1961) laudatory comments, Hennig's ideas were known best to German-speaking entomologists until the largely rewritten English version of his book.

The impact of Hennig on our perceptions of how evolution occurred and how best to analyze and represent it cannot be overemphasized, and yet as is often the case, the current state of the sciences traverses far beyond what Hennig might have imagined. As noted at its Web site, "the Hennig Society was founded in 1980 with the expressed purpose of promoting the field of Phylogenetic Systematics. Hennig's idea that groups of organisms, or taxa, should be recognized and formally named only in cases where they are evolutionarily real entities, that is 'monophyletic,' at first was controversial. It is now the prevailing approach to modern systematics." Just how profoundly Hennig's German-language book (1950) differs from the English-language book (1966) becomes clear in what he chose to illustrate. The

370-page book from 1950 has fifty-eight figures dominated by twenty maps with biogeographic distributions mostly of insects and eleven illustrations of insect anatomy, whereas only ten present one or more tree-like diagrams of some sort. By contrast, the 263-page book from 1966 has sixty-nine figures of which only seven show maps and six show anatomy, whereas forty-five present one or more tree-like diagrams. As with Darwin a century before, Hennig realized that a major key to the understanding of evolutionary history is a good knowledge of the biogeographic distributions of species. Clearly, a profound change occurred in the sixteen years between the publication of the original and the translated text, but Hennig did not diminish the role of biogeography. Rather, he expanded and developed how to probe for the underlying methodology for building and drawing tree-like diagrams. These became commonly known as cladograms by virtue of the clades, or branches (Gk. *kladoi*), that form the diagram and thus provided the simplified name for this school of systematics: cladistics.

Stated simply, cladistics tries to tease apart the similarities or characters that species and higher taxa share by virtue of a common ancestor versus those they retain from a common ancestor or those that arose through similar ecological constraints. Figure *7.1A*, found in *Phylogenetic Systematics* (1966) but not in *Grundzüge* (1950), shows the development of Hennig's ideas related to this topic. For the uninitiated, the use of lowercase and uppercase letters in figure *7.1A* at first might be confusing. The common ancestor of *A*, *B*, and *C*, which is unlabeled, possessed characters *a*, *b*, and *c*. Character *a* was transformed to *a'* only in taxon *C* so is unique to *C*. The ancestral character was retained in *A* and *B*, and thus is a shared-ancestral character, or a symplesiomorphy in Hennig's terminology, but being ancestral, such retentions are not useful in arguing evolutionary relationships between such taxa as *A* and *B*. Character *b* was transformed to *b'* in the unnamed common ancestor of *B* and *C*, and thus is a shared-derived character of *B* and *C*, or a synapomorphy in Hennig's terminology. This makes it useful in arguing that *B* and *C* share a more recent ancestor with each other than either does with *A*. Finally, *c* was transformed to *c'* separately in

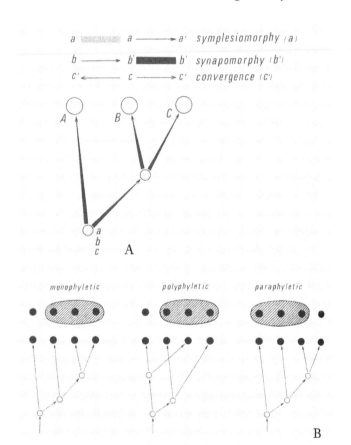

both *A* and *C*, likely as the result of similar ecological constraints, and thus they are the result of convergences. Synapomorphies and symplesiomorphies collectively represent homologies or similarities resulting from a common descent, more recent for synapomorphies and farther in the past for symplesiomorphies. Convergences caused by similar ecological constraints are often called homoplasies in juxtaposition to homologies. To establish a reasonable hypothesis of evolutionary relationships, in most real cases we need to evaluate many more than the three characters shown in Hennig's hypothetical case.

A simple example may suffice. Birds possess wings because their most recent common ancestor evolved wings; thus wings in birds are a synapomorphy for birds relative to other nonflighted vertebrates, but wings cannot be used to argue relationships among birds because they are a symplesiomorphy within Aves (birds). In contrast, the presence of wings in birds, bats, and insects occurs not because of common descent but because of the ecological constraint of the need for wings for flight—that is, a convergence (or homoplasy). Each of these three groups evolved wings from a separate ancestry, which is quite clear in this case because the underlying structures of the wings are so profoundly different. Synapmorphies form the basis for building cladograms because they arise from common descent. Hennig began developing these ideas in *Grundzüge* (1950), with a much clearer explanation in *Phylogenetic Systematics* (1966). One might ask why these ideas that appear deceptively simple were not codified earlier. In fact, a number of biologists presaged Hennig's ideas, but it was Henning and especially some of his early followers who rather strictly tried to follow the dictum that a cladogram must be built using only synapomorphies arising from the common ancestry of a given group, or, more correctly, the common ancestry for a clade. Hennig and his adherents argued further that any resulting classification must adhere to the underlying cladistic analysis. Another diagram found in *Phylogenetic Systematics* but not in the earlier *Grundzüge* shows the basis for this classification (see figure 7.1B). As Hennig labels them for the diagram, groups are monophyletic if they are based on synapomorphies, paraphyletic if based on symplesiomorphies, and polyphyletic if based on convergences.

It was and still is the issue of classification and not the philosophy of the cladistic analysis that caused the greatest consternation for biologists. Biologists can be quite conservative, not easily giving up older ways of doing things, such as the Linnaean system. An organization somewhat more sympathetic to a Hennigian classificatory system, the International Society for Phylogenetic Nomenclature, advocates the development of a classification called the Phylocode that someday may replace or at least supplement the Linnaean system. The issues become quite clear with a classic example of how to classify tetrapods, which traditionally include Amphibia, Reptilia, Aves, and Mammalia. Although the placement of Testudines (turtles) remains in flux, the three identical cladograms in figure 7.2 show the traditional view of tetrapod relationships. Using Hennigian principles, one can ask what synapomorphies traditional reptiles share. Certainly they have scales, but so do a number of mammals on their tails and birds on their legs. Feathers are, in fact, modified scales. Reptiles also share cold-bloodedness and, except for crocodilians, possess a three-chambered heart. These, however, are symplesiomorphies for Reptilia or synapomorphies of a much earlier ancestor and thus tell us nothing about relationships within Reptilia. A cladistic classification includes Aves within Reptilia, making it monophyletic, or what Hennig designated as holophyletic. This was and to a lesser extent remains anathema to some biologists. A monophyletic group or a clade (see figure 7.2A) includes the most recent common ancestor (the asterisk in figure 7.2A–C) and all its known descendants. The more traditional grouping of Reptilia forms a paraphyletic group (see figure 7.2B) because it excludes one descendant clade (Aves), although the most recent common ancestor is included. Some biologists argue that both monophyletic and paraphyletic groups should be

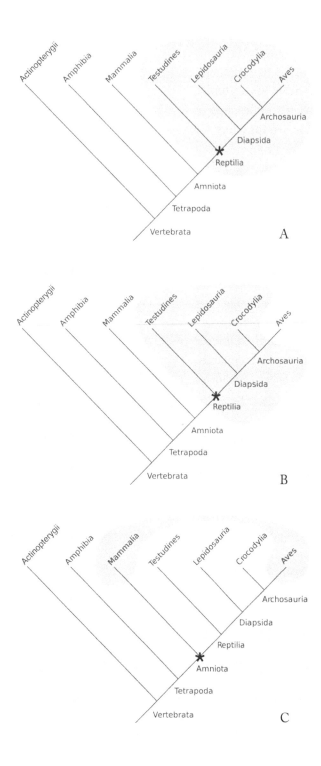

FIGURE 7.2 An example within
Vertebrata showing (A) monophyletic Reptilia, (B) paraphyletic
Reptilia, and (C) polyphyletic Haemothermia. (Adapted from various
sources)

recognized, but at most no one argues that the third
kind, called a polyphyletic group, should be recognized.
In the example in figure 7.2*C*, Haemothermia, meaning "warm-blooded," was proposed by Richard Owen
in the nineteenth century and revived for a short time
in the 1980s (Gardiner 1982). Such polyphyletic groups
have neither all known descendants nor the most recent
common ancestor.

Between Hennig's *Grundzüge* (1950) and *Phylogenetic Systematics* (1966), we see how other of his ideas
and visualizations of these ideas transformed. Like his
countryman Haeckel, Hennig utilized a variety of terms
and representations. The relationship between ontogeny
and phylogeny, terms that Haeckel coined, interested
both men. In *Grundzüge*, Hennig shows us a relatively
simple rendition based on earlier sources of what he
terms the tokogenetic relationships—that is, genetic
relationships between individuals in a species (read left
to right), specifically females and males represented
as empty and filled circles, respectively (figure 7.3*A*).
Although representing a sexually reproducing species,
the connecting lines do not present a clear delineation
of reproduction and descent. This becomes far more
explicit and clearer in *Phylogenetic Systematics*, in which
Hennig presents three figures, each more complex than
that in *Grundzüge*, explaining tokogenetic relationships.
Each succeeding diagram adds additional complexity,
with that shown in figure 7.3*B* the most complex, combining Hennig's ideas of genetic relationships of individuals in sexually reproducing species with his ideas of
transformations within the ontogeny of an individual
and the process of speciation. As an entomologist,
Hennig concerned himself more than other biologists might with the radical morphological changes
undergone by some insects from larva to pupa to adult.
He called each of these stages a semaphoront in the
metamorphosis during the ontogeny of an individual,
something of less obvious concern to, for example, an
ornithologist studying birds. These semaphoronts then
combined to form the ontogeny of a male or female, which, in turn, would mate
to form the next individual. In combinations, these repeated life cycles, matings,
and life cycles formed what Hennig termed cyclomorphism. If a barrier to breeding should form (the wedge in figure 7.3*B*), populations could split, establishing
separate breeding populations and ultimately forming two or more new species.
Hennig combined all these ideas in what he called hologenetic relationships, or
what we today usually refer to as monophyly, discussed previously, in which one

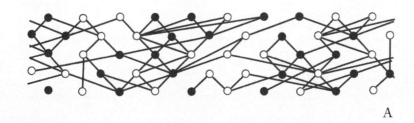

A

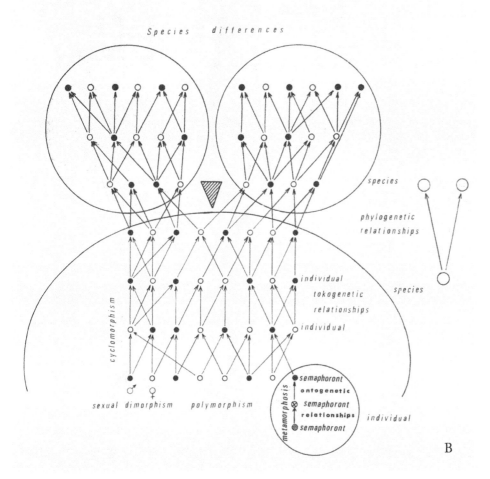

Species differences

species

phylogenetic
relationships

individual
tokogenetic
relationships

species

individual

cyclomorphism

sexual dimorphism   polymorphism

metamorphosis

● semaphoront
ontogenetic
⊗ semaphoront
relationships
◉ semaphoront

individual

B

FIGURE 7.3 Hennig's (*A*) relatively simple depiction of the tokogenetic, or genetic, relationships between individuals in a species, specifically females and males, redrawn from *Grundzüge einer Theorie der phylogenetischen Systematik* (1950), and (*B*) more explicit and clearer figure showing tokogenetic relationships, combining his ideas of genetic relationships of individuals in sexually reproducing species with his ideas of transformations within the ontogeny of an individual (semaphoronts) and the process of speciation (*wedge*), from *Phylogenetic Systematics* (1966). ([*B*] reproduced with permission of the University of Illinois Press)

or more species can be traced to a common ancestry. Hennig provides us with a small, coarser-scaled diagram to the right emphasizing the next higher level in these ongoing ontogenetic and phylogenetic processes. His ideas on recovering monophyletic relationships endure in modern systematics, whereas much of his nomenclature of these processes does not.

When Hennig began his endeavors, computing was in its infancy and thus earlier attempts at cladistic reconstructions were done by hand. The methods of representation varied, but often the sort of horizontal bars shown in figure 7.4*A* were drawn across a cladogram to show how characters transformed along the various clades. Such representations became scarcer as computing power increased and the number of characters to be analyzed rapidly expanded. The Swedish entomologist Lars Brundin (1897–1993) was one of the earliest and most influential

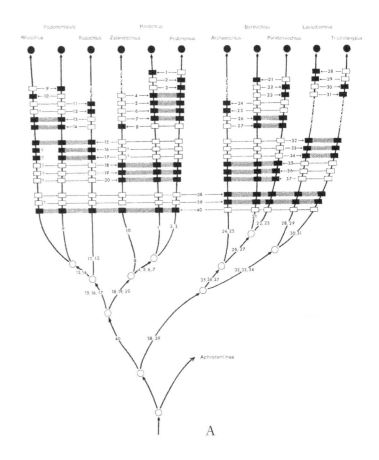

A

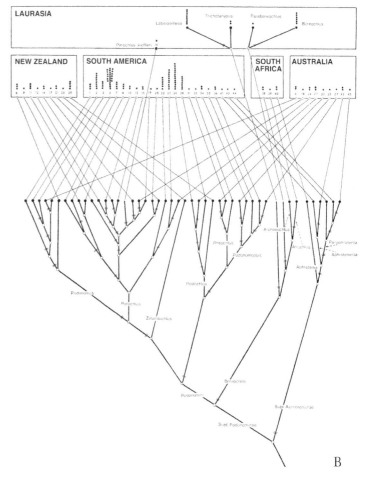

B

FIGURE 7.4 Lars Brundin's (A) cladogram and (B) area cladogram for chironomid midges, from "On the Real Nature of Transantarctic Relationships" (1965). The cladogram shows the method of representation developed by Hennig. (Reproduced with permission of the Society for the Study of Evolution)

advocates for Hennig's ideas. In 1965 (and in a longer paper in 1966), Brundin proved to be an even better spokesman for Hennig's methodology with the publication of "On the Real Nature of Transantarctic Relationships," in which he examined the phylogenetic relationships and biogeography of one group of chironomid midges. Except for entomologists, few students of evolution knew of chironomid midges, other than possibly field scientists who understood these flies to be pesky but thankfully nonbiting. Brundin provided one of the first studies in English demonstrating Hennig's methodology for not only the phylogenetic reconstruction of a specific group but also its biogeographic history using similar techniques. Brundin's cladogram follows Hennig's ideas and uses similar iconography, showing plesiomorphies as open rectangles, symplesiomorphies as open rectangles joined by a thin line, apomorphies as solid rectangles, and synapomorphies as solid rectangles joined by gray bars (see figure 7.4A). The horizontal arrows refer to specific character transformations, such as characters 38 and 39 (near the middle of the cladogram) transforming from the plesiomorphic (ancestral) states in the clade on the left to the apomorphic (derived) states in the clade on the right. The transformation of character 40 occurred in the opposite direction. Lower on the diagram, Brundin shows us where in the cladogram his analysis suggests these evolutionary transformations took place. This profoundly new and different approach that Hennig advocated and that Brundin so aptly demonstrated offered a hypothesis of relationships that could be explicitly tested. Nothing quite like it had come before. With this testable hypothesis of relationships, Brundin further demonstrated what Hennig also advocated: that such hypotheses of relationship could also provide a hypothesis of relatively when and how these various clades reached their present biogeographic ranges in what became known as area cladograms (see figure 7.4B). It must be kept in mind that both the cladogram of relationships and that of biogeographic history are hypotheses, but as we see, neither had been so explicitly and scientifically (testably) expressed before.

The second new school of "numerical taxonomy" as developed mostly through the efforts of the biologists Peter Sneath (1923–2011) and Robert Sokal (1926–2012) briefly remained at the fore for a relatively short time, basically in the 1960s and 1970s. In their two major works, *Principles of Numerical Taxonomy* (Sokal and Sneath 1963) and *Numerical Taxonomy* (Sneath and Sokal 1973), they argued that biological classifications should be based on overall similarities of species rather than on assumptions about unknowable evolutionary relationships, although they granted that when done properly, the classification to some extent reflected evolutionary history. In their earlier book, Sokal and Sneath (1963) write: "In developing the principles of numerical taxonomy, we have stressed repeatedly that phylogenetic considerations can have no part in taxonomy and in the classificatory process. Once a classification has been established, however, biologists will inevitably attempt to arrive at phylogenetic deductions from the evidence at hand" (216).

Sneath and Sokal's (1973) discussion of a figure contrasts clearly the distinction between numerical taxonomy and phylogenetic systematics (figure 7.5A). In this hypothetical phylogenetic tree, lineage x departs phenetically, or in its overall similarity, from its nearest relatives in cluster A that resemble more members of cluster B. About this the authors write, "Although the divergence of x from cluster A is of

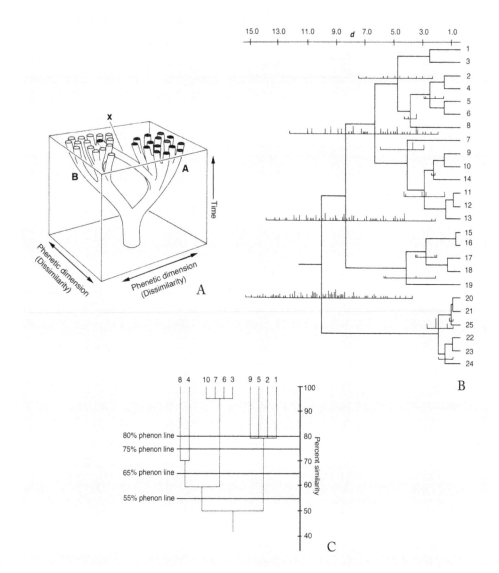

FIGURE 7.5 Peter Sneath and Robert Sokal's (A) hypothetical phylogenetic tree showing lineage x, which they would classify with cluster B rather than cluster A because of phenetic similarities to B; (B) rooted phenogram with OTU numbers on the right and frequency distributions of similarity coefficients along the horizontal axis; and (C) phenogram translated into a taxonomy using percentage-based phenons, redrawn from *Numerical Taxonomy* (1973).

great evolutionary interest, the overall similarity of x to the members of cluster B is more useful" (57), but this prompts the question as to what "useful" refers. These views were anathema to cladists, for whom the main objective or "usefulness" of a classificatory scheme was to reflect evolutionary history as nearly as possible. For them, a hypothesis of evolutionary history based on a phylogenetic analysis came before attempts at classification. Cladists' establishment of this objective occurred, as discussed earlier, by grouping taxa based on shared-derived characters or synapomorphies, which arguably reflected evolutionary history.

In numerical taxonomy, analyses of overall or phenetic traits of organisms increasingly were based on the then newly emerging computer-based algorithms from which one could obtain an arguably objective taxonomy. In addition to an increased reliance on computers, one very useful term that outlived phenetics is "operational taxonomic unit" (OTU), generally defined as "the taxonomic unit of sampling used by the researcher in a particular study." OTUs may be individual organisms, populations, species, genera, and so forth, or, for microbiologists, strains of bacteria, which are not easily classified in traditional taxonomic schemes. The results of phenetic analyses sometimes appeared as unrooted networks with distances between the OTUs indicating the degree of overall similarity. Although no longer common in building higher-level phylogenetic studies, such networks remain an important tool in population studies within a species or even very closely related species because the goal is an assessment of intertwining genetic affinities rather than phylogenetic reconstruction per se.

Sneath and Sokal (1973) referred generally to tree diagrams as dendrograms, as had others, with the further distinction for cladograms constructed by phylogenetic systematists, whereas phenograms refer to dendrograms derived from phenetic analyses. Phenograms do not represent the hypothesis of descent from a common ancestor inherent in cladograms. When used, visual representations varied but most commonly appeared as rooted phenograms that opened up, down, or, in the common case, from left to right (see figure 7.5B). In this example from a source used by Sneath and Sokal (1973), the OTUs are the numbers on the right. The authors identify the numbers at the top and the horizontal lines with small hash marks within the figure as "frequency distributions of similarity coefficients"—in other words, as a measure of overall similarity rather than any argument for recency of common ancestry. Such analyses might or might not be the same as their evolutionary counterparts. When this phenetic information translates into a taxonomy, overall similarity arguably provides the best basis for classification. One possible way is to divide a phenogram into levels or percentages of overall similarity. OTUs of a given level or percentage of similarity could be thus grouped, such as the so-called percentage-based phenons (see figure 7.5C). This presents a good example of the push for objectivity leading instead to arbitrariness in taxonomic classification.

Although unquestionably accepting evolution, Sneath and Sokal argued that recovery of its history was difficult or impossible to accomplish. They claimed that when done properly, a phenetic classification to some extent reflected parts of evolutionary history. The question remains: What parts of evolutionary history are reflected under the aegis of phenetics? Even though phenograms and cladograms superficially resemble each other in being tree-like, just as with the nonevolutionary and evolutionary trees of the nineteenth century, the intentions of the two diverge quite markedly from each other. Once again, very similar iconography meant something quite different, depending on the underlying theoretical framework used to generate it. In the end, phylogenetic systematics/cladistics took hold while numerical taxonomy/phenetics faded, as measured by the number of books that the former school rather quickly spawned. Books espousing phylogenetic systematics/cladistics appeared soon after the reprinting in 1979 of Hennig's *Phylogenetic Systematics* (1966) (for example, Eldredge and Cracraft 1980; Nelson and Platnick 1981; Wiley 1981), but not those promoting numerical taxonomy/phenetics following the publication of Sneath and Sokal's *Numerical Taxonomy* (1973). The rise of both molecular tools and far more powerful computing certainly accelerated this shift, but, more important, an underlying desire drove it: the need to understand how species relate one to another and, most important, how humans relate to other species. Cladistics strove to provide the answer, whereas phenetics did not.

The Molecular Approach to Tree Building

At the same time as the schools of numerical taxonomy/phenetics and phylogenetic systematics/cladistics emerged with more objective methods of building trees and creating classifications, other new technologies further revolutionized the reconstruction of life's history. These new, molecularly based technologies allowed us to explore in far greater detail and depth how cells function, how proteins are synthesized, and how genetic information is passed from one generation to the next. As with anatomical characters, molecularly based characters varied in what they represented. Recall that in the latter half of the nineteenth century, St. George Jackson Mivart (1865, 1867) used various aspects of primate postcranial anatomy to explore primate relationships, knowing full well that these characters alone could not provide a complete portrait of primate evolution. In the mid-twentieth century, William King Gregory had taken this approach to a greater and more integrated scale in the study of various organ systems and how they integrate to form the whole—an advance that proved very useful in phylogenetic reconstruction. So, too, when molecularly based analyses became possible, they began by using more clearly expressed cellular-level functions and those that could relatively easily be analyzed across various species. Thus the earliest attempts at molecular systematics did not look directly at the basis of genetics—deoxyribonucleic acid (DNA) and ribonucleic acid (RNA)—but indirectly at the products of these molecules: proteins and their building blocks, amino acids. Other early attempts were even more indirect, using the strength of immunological responses to infer closeness of relationship. In these early stages of molecular systematics, little concern arose as to which school of systematics should be used, although most would fall within the purview of phenetics because they examined overall similarity.

Tree-like diagrams using molecular or genetic data began to appear in earnest in the early 1960s, some ten years after the discovery and description of DNA in the early 1950s. The truly embryonic stages of molecularly based analyses, quite surprisingly, occurred far earlier than usually recognized, with the first attempts at molecular systematics dating to the very beginning of the twentieth century (Wood 2012). In 1901, the American British bacteriologist George H. F. Nuttall (1862–1937) reported comparing upward of 230 blood samples for all classes of vertebrates. In 1902, the work of British physiologist Albert S. F. Grünbaum (1869–1921, name changed to Leyton in 1915) with blood serums showed the great similarity among humans, chimpanzees, gorillas, and orangutans, arguably making him the pioneer of comparative primate serology (Wood 2012). Nuttall wrote additional papers on the topic with other colleagues, producing in 1904 an impressive 444-page monograph in which he demonstrates "certain blood-relationships amongst animals as indicated by 16,000 tests made by myself with precipitating antisera upon 900 specimens of blood obtained from various sources" (vii). In this monograph, Nuttall did not provide a phylogenetic tree based on his results, but he did include a tree about which he notes: "According to Dubois (1896) the relationships amongst the Anthropoidea are represented by the accompanying genealogical tree, based upon that of Haeckel (1895)" (figure 7.6A). He goes on to describe Eugène Dubois's ideas on the fossil species included on the diagram, following

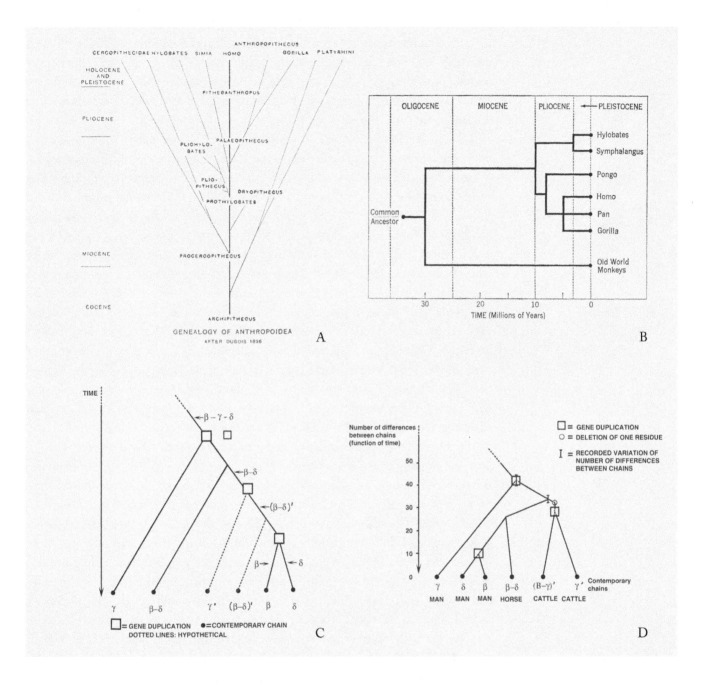

FIGURE 7.6 (*A*) Phylogenetic tree, modified after one in Eugène Dubois's "*Pithecanthropus erectus,*
eine Stammform des Menschen" (1896), showing relationships among Anthropoidea that agree
with the results from George Nuttall's blood serum results, from *Blood Immunity and Blood Relationship*
(1904); (*B*) Vincent Sarich and Allan Wilson's tree for primate evolution, showing an unresolved
tripartite relationship among humans, chimps, and gorillas with a quite recent timing for this split,
from "Immunological Time Scale for Hominid Evolution" (1967); Emile Zuckerkandl and Linus Pauling's
downward-opening trees, showing (*C*) the process of gene duplication and (*D*) the evolutionary relationships of some mammalian hemoglobin chains relative to the species in which they occur, modified
from "Evolutionary Divergence and Convergence in Proteins" (1965). ([*B*] Reproduced with permission
of the American Association for the Advancement of Science)

which he continues, "the fact remains, that the degrees of reaction obtained by me in my blood tests are in strict accord with this genealogy, as pointing to the more remote relationship of the Cercopithecidae, but especially of the New World monkeys" (2). The relationships as shown do not differ greatly from those presented some sixty years later and accepted today, except that now chimpanzees ("Anthropopithecus") are grouped with humans rather than with gorillas, a change not suggested until the early 1960s.

The early 1960s proved pivotal in the publication of papers using molecular techniques to produce phylogenies; some were more theoretical, whereas others presented specific results. In 1965, the supposedly first phylogeny using the principle of parsimony was published by Luigi Cavalli-Sforza and Anthony Edwards, showing relationships of humans populations based on blood-group polymorphism gene frequencies (Pietsch 2012)—that is, the relative frequencies of the blood types A, B, AB, and O in human populations. In 1964, a paper written in English by Emile Zuckerkandl (1922–2013) and Linus Pauling (1901–1994) was translated and published in Russian; in 1965, it was published in English with the intriguing title "Molecules as Documents of Evolutionary History." Although the paper lacks a tree figure, Zuckerkandl and Pauling (1965b) noted in the very first sentence of the abstract: "Different types of molecules are discussed in relation to their fitness for providing the basis for a molecular phylogeny." They continued that "semantides," a term they coined for "the different types of macromolecules that carry the genetic information or a very extensive translation thereof" (rRNA, rDNA, and so on), would work best for such research (357).

In a book chapter, Zuckerkandl and Pauling (1965a) discussed the evolution of proteins, notably of the mammalian hemoglobin chains—that is, variations in the subunits of hemoglobin, the protein that transports oxygen in the red blood cells of vertebrates. The paper includes two downward-opening diagrams, the first of which deals with the process of gene duplication yielding what they called β (beta), γ (gamma), and δ (delta) chains of hemoglobin, clearly providing a phylogeny of this protein rather than of any organisms (see figure 7.6C). The second figure, also downward opening, at first appears to show a phylogeny of humans, horses, and cattle using the same protein (see figure 7.6D). It would be an odd phylogeny indeed that has horses sharing a closer ancestry with some humans than these humans share with other humans—shades of the centaur? It becomes clear in both the labeling of the diagram and the discussion that this study analyzes the probable evolutionary changes of some mammalian hemoglobin chains rather than the relationships of the species in which these hemoglobin chains occur. Nonetheless, Zuckerkandl and Pauling note that results such as these offer an intriguing possibility for doing phylogenetic studies but that the evidence alone cannot be taken seriously, as it relies on a single protein—again shades of Mivart's late-nineteenth-century caution about using single organ systems to deduce relationships. How gene families evolve remains a rich area of study along with the potentially vexing issue of how to relate gene trees with the evolutionary history of the organisms that contain them, inasmuch as the two sometimes are not congruent. These results offer an early warning that the evolutionary history of molecules does not always neatly track one for one the history of the species in which they occur.

In the same chapter, Zuckerkandl and Pauling (1965a) described a hypothesis that they had been promoting for the previous three years, which they called the molecular evolutionary clock (Morgan 1998). In one form, the hypothesis argues that the amino acids composing a given protein (or other noncoding parts of the genome), not being under selective pressure, change in a stochastically constant manner over time, thus approximating clock-like change. In theory, then, this "clock" can be used to estimate the time since two given species or other groups split one from the other; fossils need not apply to establish this event, but as we shall see, this has become a major area of contention between molecularly based and anatomically based systematists concerning the timing of some major evolutionary events. Although Zuckerkandl and Pauling's chapter helped set the stage for what followed, its most frequent citation comes not because of the early molecular trees discussed earlier but for the molecular clock hypothesis.

A second molecular technique championed in the 1960s in the laboratory of Allan C. Wilson (1934–1991) became known as the microcomplement fixation method. The technique relied on the fact that an organism's body builds antibodies in an immunological response to foreign substances termed antigens—in this case, between various proteins. The closer two species relate to each other, the stronger the antigen–antibody response, such as between a chimpanzee and a human versus either of these species and a species of baboon. Assumptions for this analysis rely on the hypotheses that the differences seen between species relate to the amount of changes in amino acids within the proteins and, more controversially, that these changes follow some sort of molecular clock model such as that advocated by Zuckerkandl and Pauling (1965a, 1965b). Also, the technique measures overall similarity such as that advocated by numerical taxonomy.

Beginning in the 1960s, trees using this technique appeared, notably in a paper published in 1967 by Vincent Sarich (1934–2012) and Wilson, the mentor for his doctoral degree, dealing with primate evolution (see figure 7.6B). Showing an unresolved tripartite relationship among humans, chimps, and gorillas caused some controversy, but the quite recent timing of the split created a much greater yelp from critics. Sarich and Wilson arrived at a date of about 5 million years ago for this three-way split based on immunological distances, contrasting greatly with estimates of up to 30 million years based mostly on the fossil record. Current estimates based on fossils and molecules bracket this possibly complex speciation process between chimps and humans from 10 to 5 million years ago. More to the point, whereas Zuckerkandl and Pauling's (1965a, 1965b) papers created smoke over the issue of a molecular clock, Sarich and Wilson's (1967) paper set the place ablaze. Although both sides have given some ground—there is not one but many molecular clocks, and molecular methods sometimes resolve relationships not possible using fossils—the greatest issue remains the quandary over the timing of phylogenetic events and whether fossils, molecules, or both best address and answer these questions.

A paper by Morris Goodman (1925–2010) appears to predate Sarich and Wilson (1967) in arguing for the tripartite relationship among humans, chimps, and gorillas. Although not as explicitly, Goodman also suggested that some molecular changes are neutral relative to natural selection and hence could change in a stochastically random fashion. In 1963, Goodman published what he called a primate "classi-fication from serological reactions" (that is, antigen–antibody reactions between blood serums of various mammalian species), although in the text he does refer to it as a dendrogram (figure 7.7A). While a classification is certainly part of this figure, it equally shows Goodman's ideas of relationships among various primates, with other mammals as outliers. Particularly interesting, Goodman shows gorillas, chimpanzees, and humans as an unresolved tritomy, or three-way split, as Sarich and Wilson would do four years later. Further, Goodman rather audaciously places all three into the "human family," Hominidae. Recall that although earlier workers well back into the nineteenth century, Darwin among them, argued for a special relationship between apes and humans, they never suggested placing them in the same family. Even in the middle of the twentieth century, to distinguish a special relationship among gorillas, chimpanzees, and humans to the exclusion of other apes was going too far for many, if not most, biologists. But unlike in preceding centuries, this had less to do with any obeisance to an Abrahamic God than with the quite obvious physical, behavioral, and intellectual differences separating humans from apes.

One person in particular at the meeting in 1962 at which Goodman presented his ideas of grouping gorillas, chimpanzees, and humans within Hominidae was George Gaylord Simpson (1963), who presented a paper for the volume in which Goodman's paper also appeared (Washburn 1963). In his accompanying phylogeny, Simpson shows the traditional phylogenetic tree placing humans (*Homo*) to the right as quite distinct from the apes shown to the left (see figure 7.7B). Notably, Simpson includes chimpanzees and gorillas in his genus *Pan*. His classification later in the paper has only *Homo* and *Australopithecus* in Hominidae, with all but a few odd fossils placed in the ape family Pongidae. Simpson was not simply being obstreperous; recall that he, as did most supporters of the Modern Synthesis in the mid-twentieth century, championed Darwin's adaptational changes and wished to reflect them not only in phylogenies but in classifications as well. Figure 7.7C clearly presents Simpson's (1963) arguments that humans represent such a differ-ent ecological milieu from apes that they must be placed in a different "hominid zone." As in figure 7.7B, *Pan* includes both gorillas and chimpanzees. Once again, Simpson is using a tree to argue his case not only for the pattern of evolution but also for its process. For Simpson, adaptational regimes play an important role in phylogenetic reconstruction along with classification.

FIGURE 7.7 (*A*) Morris Goodman's classification, or dendrogram, of primates based on serological re-actions, placing gorillas, chimpanzees, and humans into the "human family," Hominidae, modified from "Man's Place in the Phylogeny of the Primates as Reflected in Serum Proteins" (1963); George Gaylord Simpson's (*B*) traditional phylogenetic tree, showing humans as quite distinct from the apes and plac-ing chimpanzees and gorillas in the genus *Pan*, and (*C*) figure arguing that humans must be placed in a different "hominid zone" from apes, modified from "The Meaning of Taxonomic Statements" (1963); (*D*) Goodman, G. William Moore, and Genji Matsuda's reconstruction of the genealogy of the globin chains, from plants to humans, from "Darwinian Evolution in the Genealogy of Haemoglobin" (1975). ([*D*] reproduced with permission of the Nature Publishing Group)

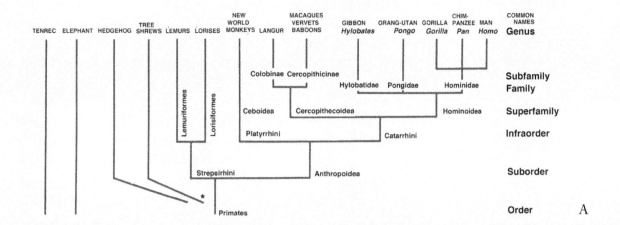

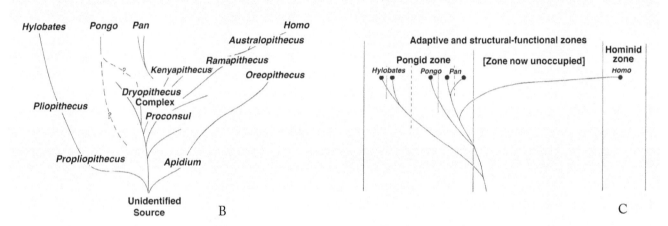

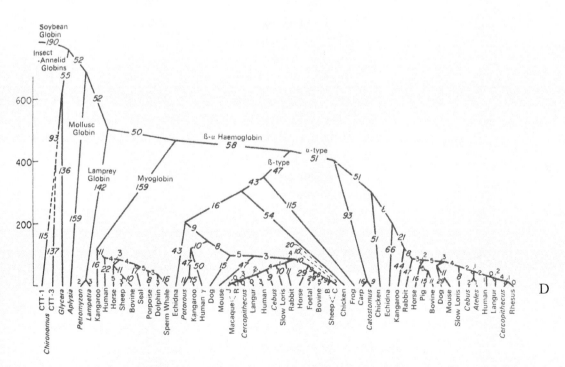

Goodman had begun his advanced academic training with work in the emerging field of molecular biology by notably dealing with immunological work on hemoglobin. Because of his interest in evolutionary questions, in the late 1950s he gravitated to looking more widely at differences among other mammalian blood sera, often obtained from the Brookfield Zoo near Chicago. Goodman became an undoubted leader in molecular systematics especially later in his career, but early on, although work on molecular systematics was a passing fancy for some notables in molecular studies (that is, its importance was lessened because it did not deal with curing human diseases), molecular systematics was central to Goodman's research program. On the opposite end, morphologically based systematists showed less than a keen interest in this new-fangled approach. As Goodman expressed in an interview (Hagen 2004), "In biomedical research it would be somewhat unusual just to be spending all afternoon thinking about something," by which he surely meant the study of evolutionary history using molecular techniques. Goodman acknowledged the much earlier but nonetheless important contributions of Nuttall at the beginning of the twentieth century, realizing that unlike Goodman, Nuttall had little information to understand the genetic basis for his phylogenetic work. Goodman expressed well what many scientists struggled with in the late 1950s and early 1960s: "So, if I were to criticize myself at all . . . the main criticism I would have is that I had more or less accepted the thought that there is progressive evolution from less organized to more organized and more complex, and that humans are at some sort of pinnacle. I don't think I have this view anymore." Humans finally could be seen as part of nature, rather than above it. Goodman clearly bridged the various areas and levels of biology:

> When it comes to the bigger questions we should not view molecular evolution as divorced from organismal evolution. I think there are tremendous interconnections among these levels of evolution. This of course is an area of biological science that tries to emphasize the integration of the different levels. The reason we think for reconstructing phylogeny that molecular data is so important is because of things such as Simpson said that it would be priceless data to be able to map the stream of heredity. That's exactly what the genome sequencing is allowing. But it's only in its infancy in terms of the amount of genome data that we need to adequately map the stream of heredity. I think it's moving along with the primates pretty well. It probably will influence the nature of biology for the next 50 years or so. (quoted in Hagen 2004)

Goodman not only used antigen–antibody seriological studies for his work but later began to use sequence data of mutations in various proteins to build phylogenetic trees of both organisms and the genes they contain. An early paper reconstructed the genealogy of the globin chains (Goodman, Moore, and Matsuda 1975)—that is, the protein components of hemoglobin and myoglobin found, respectively, in red blood cells and muscles cells of vertebrates (see figure 7.7D). In this phylogeny of globins, note that humans show up four times because we possess varieties of myoglobins, as well as α and β chains of hemoglobin. Other evolutionarily related globins shown come from much more distantly related plants and nonvertebrates. An important hypothesis of these authors, as indicated by their paper's title, "Darwinian Evolution in the Genealogy of Haemoglobin," was that the rates of evolution of these globins could be explained not "by the theory of random fixation of selection, so-called non-Darwinian evolution" inherent in the idea of a universal clock for molecular change, but by its attribution "to positive selection for more optimal function" (603). Whether correct or not, as with other prior notable biologists such as Darwin and Simpson, Goodman and his colleagues used trees to argue aspects of the evolutionary process and not simply evolutionary pattern.

Charles Gald Sibley (1917–1998), as with Goodman, was particularly interested in forging new molecular techniques for use in systematics. Sibley's approach was quite different, developing a technique known as DNA–DNA hybridization. This decidedly laborious process required stamina and tenacity as much as most fieldwork, which Sibley undertook early in his career in the South Pacific and elsewhere. Sibley's earlier research examined the process of hybridization between closely related species or subspecies of birds, notably using the variations in the proteins found in the egg albumin of these birds. In the late 1950s, Sibley and his colleagues began using a process called electrophoresis, first with a paper medium and later with a gel, in which a small electrical current passed through the medium causes the protein molecules to spread out according to charge and size, which then could be interpreted to determine relative differences among the included species or subspecies. Fortuitously, it also revealed discernible differences among more distantly related species (Ahlquist 1999). Sibley and Paul Johnsgard (1959a, 1959b) published two papers dealing with variations in protein electrophoretic patterns of various avian species. While not presenting a phylogenetic tree, Sibley and Johnsgard (1959b) made it quite clear "that particular proteins characterize every species of plant and animal and that phylogenetic relationships are reflected in protein structure" (85).

Later Sibley, in an extended collaboration with Jon Edward Ahlquist (b. 1944), shifted to using DNA–DNA hybridization to contribute greatly in the late twentieth century to bird systematics, especially to some knotty problems dealing with higher relations of various bird groups. As one biographer noted, "Charles's intellectual intensity and excitement touched the lives of many of his contemporaries in ways both good and bad, and he influenced several generations of students. Few ornithologists have so polarized their students and colleagues. . . . He was a generous person, giving freely and frequently of his time to students and colleagues, particularly if it involved discussions of science" (Brush 2003:3). I can testify to this assessment of Sibley's strengths and foibles when he generously allowed me, a young paleobiologist curious about and eager to understand molecular systematics, to learn his methods in the late 1970s and early 1980s at Yale.

DNA–DNA hybridization arguably allowed better comparisons of species by measuring more direct similarities and differences between nuclear DNA of target species, although not by directly comparing sequences but by comparing the strength of chemical bonds between the DNA of species of interest. In a series of laboratory steps, first comes the purification of so-called single-copy DNA present as one copy in the haploid or sex cell. Single strands of these single-copy DNAs are then paired with single strands of single-copy DNAs of other species being sampled to create hybrid double-stranded DNA of double helix fame. When subjected to heat, the hybrid double-stranded DNAs of more closely related species with a greater number of DNA similarities remain intact at higher temperatures, whereas the hybrid double strands of more distantly related species come apart at lower temperatures, indicating that they possess fewer hydrogen bonds holding the base pairs of each DNA strand together. Phylogenies of these overall levels of similarity can thus be built with these results. Methodological problems dogged this technique, not least of which the aforementioned molecular clock assumed to be correct by Sibley, although Ahlquist (1999) indicates that Sibley never believed in uniform rates of genomic change in birds. Nonetheless, DNA–DNA hybridization yielded a new and sometimes radical phylogenetic analysis of birds of the world (Sibley, Ahlquist, and Monroe 1988; Sibley and Ahlquist 1990). Figure 7.8 shows the five tree figures from Sibley, Ahlquist, and Burt L. Monroe's (1988) paper: all extant bird clades (see figure 7.8A); all nonpasserine, or nonperching, birds (see figure 7.8B); and passerine, or perching, birds (see figure 7.8C). In this figure, each split has two numbers, the first indicating the average delta $T_{50}H$ value and the second indicating the number of hybrids that were formed between taxa on each side of split. Delta $T_{50}H$ was the somewhat controversial measure of differences that Sibley and Ahlquist developed to measure genetic divergences between various species to be compared. $T_{50}H$ represents the temperature in degrees Celsius at which 50 percent hybridization of single-copy DNA of two different species remains intact and 50 percent has disassociated. Delta $T_{50}H$ is the difference in degrees Celsius of $T_{50}H$ for hybridization of single-copy DNA within the same species versus the $T_{50}H$ for hybridization of single-copy DNA between the two species being compared. Arguably the lower the delta $T_{50}H$, the closer two given species are related to each other.

Sibley and Ahlquist's research on human ancestry using DNA–DNA hybridization proved equally controversial. Sibley and Ahlquist (1984, 1987), using DNA–DNA hybridization, provided the strongest molecular data to date that both species of chimpanzee share a more recent ancestry with humans than chimpanzees do with gorillas. This certainly presented a step beyond that of Goodman (1963) and Sarich and Wilson (1967), who linked gorillas, chimpanzees, and humans in a tritomy. In their paper from 1984, Sibley and Ahlquist not only presented their results showing humans sharing a closer ancestry with chimpanzees (figure 7.9A), but also created a tree from the work of Bill Hoyer and colleagues (1972), who had argued for the same relationships without providing a tree (see figure 7.9B). The tree in Sibley and Ahlquist's paper from 1987 differs little from their earlier tree, other than its noticeably increased sample sizes (N) (see figure 7.9C). Note that in addition to a Delta $T_{50}H$ scale, for their 1984 tree they provide a time scale in millions of years ago (mya) to indicate their assessment of the timing of the various splits of lineages, with that between humans and chimpanzees at 7.7 to 6.3 mya

FIGURE 7.8 Charles Sibley, Jon Ahlquist, and Burt Monroe's five trees based on DNA–DNA hybridization results: (A) all extant bird clades; (B) all nonpasserine birds (a combination of two trees); and (C) passerine birds, from "A Classification of the Living Birds of the World Based on DNA–DNA Hybridization Studies" (1988). (Reproduced with permission of the American Ornithologist's Union)

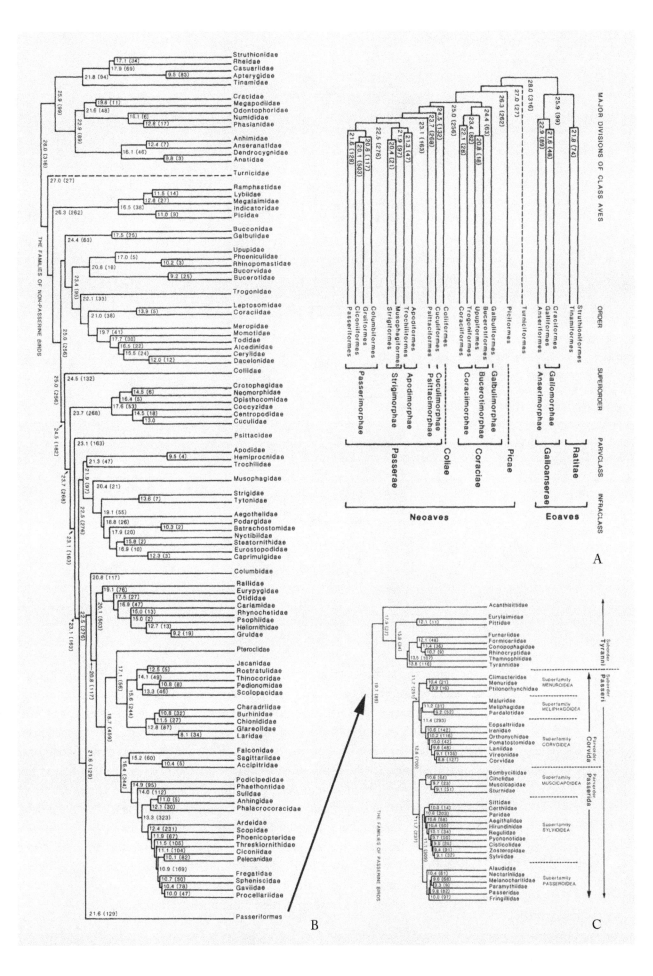

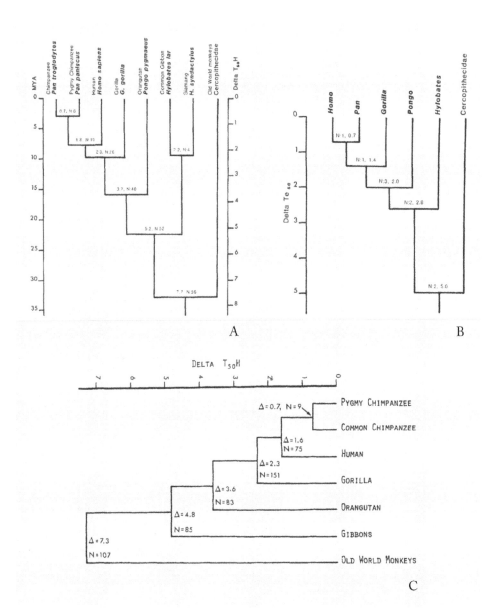

FIGURE 7.9 Sibley and Ahlquist
(A) used DNA–DNA hybridization
to argue that humans share a close
ancestry with chimpanzees and
(B) created a tree from the results
of Bill Hoyer and his colleagues,
who had argued in "Examination
of Hominid Evolution by DNA
Sequence Homology" (1972) for
the same relationships, from "The
Phylogeny of the Hominoid Primates, as Indicated by DNA–DNA
Hybridization" (1984); (C) their
later results differ little, other than
they noticeably increased their
sample sizes (N), from "DNA Hybridization Evidence of Hominid
Phylogeny" (1987). (Reproduced
with permission of Springer)

(see figure 7.9*A*). They used an estimated calibration of the origin of the orangutan clade of 16 mya based on fossils to help determine the various splitting dates.

For a variety of reasons, most notably newer molecular techniques, for DNA–DNA hybridization the "window of opportunity for significant research was from 1974 to 1986, the years of active work using DNA–DNA hybridization" (Ahlquist 1999:116). Newer results using more sophisticated methods generally support the basic Sibley and Ahlquist argument regarding the recency of common ancestry between chimpanzees and humans to the exclusion of other apes, but the results for their even more far-reaching bird work are more mixed. Even as these newer molecular techniques came to the fore from the 1960s through the early 1980s, not only did they provide newer methods of building trees, but some of them began to cast considerable doubt on the very idea that life's phylogeny might resemble a stately tree. Rather, life's history seemed more to suggest a tangled web.

Crunching Data

In spite of or possibly because of our better realization of just how much we do not know about the history of life, newer studies appear in press and online at an ever-accelerating pace. Certainly our innate curiosity as a species in part drives this rush for results, but an even more substantial reason rests with emerging technologies: we possess the capability to do it because computing power has increased by leaps and bounds along with the development of newer molecular techniques.

The first part of this equation, computing power, began to change slowly at first but with increasing speed in the late 1980s and more so in the 1990s as access, cost, and power of computers greatly expanded, essentially following the so-called Moore's law, which states that the number of transistors on a chip doubles approximately every one and a half to two years. For anyone using computers, especially desktop or laptop models beginning in the late 1970s and early 1980s, the changes have been mind-boggling: a typical smart phone is about one-hundredth the weight and one–five hundredth the volume of a portable computer of the early 1980s, yet it costs much less and its processor is one hundred times faster.

The second part of the accelerated pace for newer kinds of phylogenetic studies began to come to the fore after 1983 when Kary Mullis (b. 1944) developed PCR technology, which revolutionized all molecularly based research, including systematics. Before this, the collection, purification, amplification, and analysis of much genomic information were exceedingly expensive and time consuming. PCR changed this by permitting the amplification of only a few copies of a piece of DNA by orders of magnitude, creating millions of copies of the DNA sequence of interest. What once presented very laborious laboratory work became increasingly automated as more sequences could be generated, read, and analyzed by computer-driven software. Whether this democratization by virtue of availability helped to improve the quality of phylogenetic studies remains to be judged in the long run, but for now at least it presents a boon in our ability to build ever more complex phylogenies. Such studies coming solely from molecular data emerged as the field of phylogenomics (Pennisi 2008).

An example of the results of the newer phylogenomic approaches and technology in the first decade of this century provides a nice contrast with the bird work of Sibley, Ahlquist, and others in the final decade of the twentieth century. A phenomenon of "big" science papers, as in physics, for some time included multiple authors, something new to phylogenetic studies, such as the phylogenomic study of birds published in the journal *Science* with a total of eighteen authors (Hackett et al. 2008). This trend began almost a decade earlier, as more labs and more contributors were necessary to pull together and analyze the burgeoning data. Shannon Hackett and her colleagues examined about 32,000 DNA nucleotide bases from nineteen separate locations on nuclear DNA sequences. Their study included 169 species from all major bird groups.

Although Hackett and colleagues (2008) published a very detailed cladogram of their results, even more fascinating is the phylogeny comparing their results with those of three other major studies. Although originally in color, the grayscale version in figure 7.10 still retains enough differences to make some interesting comparisons. The phylogeny shows their results conveniently with the common names of the bird groups to the right. The three middle columns show the results

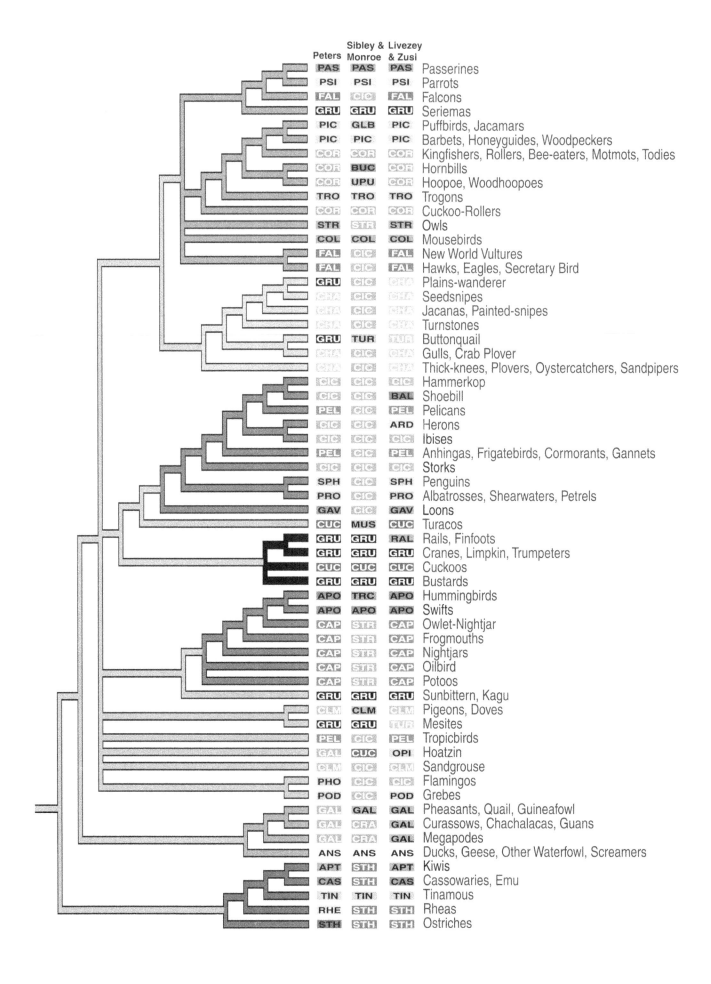

	Sibley &	Livezey	
	Monroe	& Zusi	
Peters			
PAS	PAS	PAS	Passerines
PSI	PSI	PSI	Parrots
FAL	CIC	FAL	Falcons
GRU	GRU	GRU	Seriemas
PIC	GLB	PIC	Puffbirds, Jacamars
PIC	PIC	PIC	Barbets, Honeyguides, Woodpeckers
COR	COR	COR	Kingfishers, Rollers, Bee-eaters, Motmots, Todies
COR	BUC	COR	Hornbills
COR	UPU	COR	Hoopoe, Woodhoopoes
TRO	TRO	TRO	Trogons
COR	COR	COR	Cuckoo-Rollers
STR	STR	STR	Owls
COL	COL	COL	Mousebirds
FAL	CIC	FAL	New World Vultures
FAL	CIC	FAL	Hawks, Eagles, Secretary Bird
GRU	CIC	CHA	Plains-wanderer
CHA	CIC	CHA	Seedsnipes
CHA	CIC	CHA	Jacanas, Painted-snipes
CHA	CIC	CHA	Turnstones
GRU	TUR	TUR	Buttonquail
CHA	CIC	CHA	Gulls, Crab Plover
CHA	CIC	CHA	Thick-knees, Plovers, Oystercatchers, Sandpipers
CIC	CIC	CIC	Hammerkop
CIC	CIC	BAL	Shoebill
PEL	CIC	PEL	Pelicans
CIC	CIC	ARD	Herons
CIC	CIC	CIC	Ibises
PEL	CIC	PEL	Anhingas, Frigatebirds, Cormorants, Gannets
CIC	CIC	CIC	Storks
SPH	CIC	SPH	Penguins
PRO	CIC	PRO	Albatrosses, Shearwaters, Petrels
GAV	CIC	GAV	Loons
CUC	MUS	CUC	Turacos
GRU	GRU	RAL	Rails, Finfoots
GRU	GRU	GRU	Cranes, Limpkin, Trumpeters
CUC	CUC	CUC	Cuckoos
GRU	GRU	GRU	Bustards
APO	TRC	APO	Hummingbirds
APO	APO	APO	Swifts
CAP	STR	CAP	Owlet-Nightjar
CAP	STR	CAP	Frogmouths
CAP	STR	CAP	Nightjars
CAP	STR	CAP	Oilbird
CAP	STR	CAP	Potoos
GRU	GRU	GRU	Sunbittern, Kagu
CLM	CLM	CLM	Pigeons, Doves
GRU	GRU	TUR	Mesites
PEL	CIC	PEL	Tropicbirds
GAL	CUC	OPI	Hoatzin
CLM	CIC	CLM	Sandgrouse
PHO	CIC	CIC	Flamingos
POD	CIC	POD	Grebes
GAL	GAL	GAL	Pheasants, Quail, Guineafowl
GAL	CRA	GAL	Curassows, Chachalacas, Guans
GAL	CRA	GAL	Megapodes
ANS	ANS	ANS	Ducks, Geese, Other Waterfowl, Screamers
APT	STH	APT	Kiwis
CAS	STH	CAS	Cassowaries, Emu
TIN	TIN	TIN	Tinamous
RHE	STH	STH	Rheas
STH	STH	STH	Ostriches

of three earlier studies, including the one in the middle by Sibley and Monroe (1990), quite similar to the aforementioned work of Sibley, Ahlquist, and Monroe (1988). The three letters in the rectangles indicate the orders into which each of the three other studies placed the groups named to the right. Black letters indicate that Hackett and her colleagues found the group to be monophyletic, whereas white letters indicate nonmonophyly in their study (that is, a group that they did not find shared an ancestor, to the exclusion of other groups).

Two examples show how the ideas of phylogeny changed between the earlier studies of Sibley and colleagues (Sibley, Ahlquist, and Monroe 1988; Sibley and Monroe 1990) and those of Hackett and co-workers (2008) almost twenty years later. Ciconiiforms represent the first example, shown under Sibley and Monroe (1990) as a frequently repeated white CIC. Traditionally, this group included storks, herons, egrets, spoonbills, and ibises—generally large-billed, long-legged, larger wading birds. But to these, Sibley and his colleagues added a number of other groups, which because of their scattered positions on Hackett and her colleagues' phylogeny obviously could not be monophyletic. One of the more curious ciconiiform cases that received some press attention involves New World vultures, traditionally placed with Old World vultures, hawks, and eagles, but instead Sibley and his colleagues found them to be in their greatly expanded ciconiiform order. As can be seen in figure 7.10, Hackett and her colleagues returned them to a position as being the nearest relatives to the hawks, eagles, and Old World vultures.

The second example involves the globally distributed flightless ratites—ostriches in Africa, kiwis in New Zealand, cassowaries and emus in Australia, and rheas in South America. Their monophyly has been questioned at times, but Sibley and his colleagues found them to be monophyletic with their nearest relatives, the flighted tinamous of South America. This would suggest that flightlessness originated once, and then ratites, both extant and extinct, distributed themselves around the southern continents as the southern supercontinent Gondwana broke apart. The findings of Hackett and her colleagues strongly questioned this result and in the process provided a case in which tree building showed not only how species are related but also some important evolutionary steps that occurred because of their new analysis. As seen in the base of their phylogeny, the flighted tinamous share a more recent ancestry with the common ancestor of New Zealand kiwis and Australian cassowaries and emus (see figure 7.10). They do not indicate whether they think their phylogeny argues that the ability for flight reemerged in tinamous or was lost three times, once each in the ostriches, rheas, and combined Australian–New Zealand clade. In another paper with eighteen authors (Harshman et al. 2008), most of whom also contributed to Hackett and colleagues (2008), the conclusions present a more emphatic result, as indicated by the title, "Phylogenomic Evidence

FIGURE 7.10 Results of a cladistic analysis of birds by Shannon Hackett and her colleagues, compared with the results of three earlier studies, from "A Phylogenomic Study of Birds Reveals Their Evolutionary History" (2008). (Reproduced with permission of the American Association for the Advancement of Science)

for Multiple Losses of Flight in Ratite Birds," with a major conclusion that "the most plausible hypothesis requires at least three losses of flight and explains the many morphological and behavioral similarities among ratites by parallel or convergent evolution" (13462) One can imagine that this did not make morphologists and behaviorists happy. This is but one example showing the importance of tree building in assessing evolutionary processes and not just patterns, especially following on earlier important, cladistically based biogeographers such as Brundin (1965, 1966).

The issues surrounding phylogenetic trees of birds pertain not only to who is related to whom, but to the times when various groups of birds originated. This brings us back to the arguments about the reliability and universality of the molecular clock, which, as discussed, appeared in a paper by Zuckerkandl and Pauling (1965b). Whereas arguments for the universality of rates of change of the molecular clock waned after 1965, arguments for some sort of so-called relaxed molecular clock became stronger, to the point that a clear disconnect arose between molecularly and anatomically based phylogenetic analyses and the resulting trees. Over time, more and more of a consensus arose between anatomy and molecules regarding the nature of bird relationships, but when the splitting of lineages occurred remains at an impasse, with most anatomically based phylogenies claiming that most orders of modern birds arose after the extinction of nonavian dinosaurs at the end of the Cretaceous period, some 66 mya. For example, Alan Feduccia (1995, 1996) argued that Neornithes, or modern birds, did not originate and radiate until after the Cretaceous–Paleogene (K–Pg) boundary. In what must be one of the more imaginative phylogenetic diagrams (figure 7.11*A*), Feduccia shows not only the origin and radiation of Neornithes after the K–Pg boundary but also a very substantial radiation and extinction of Enantiornithes (Archibald 2011). These are a curious lot; in fact, the name means "opposite birds." Although enantiornithines certainly appear to have been ecologically diverse in the Late Mesozoic, there are only thirty named species, and to my knowledge there is only one named species from the latest Cretaceous, although two other unnamed forms occur (Longrich, Tokaryk, and Field 2011). This is a far cry from the radiation in Feduccia's phylogenetic diagram that makes them a close second to neornithines, which are known from more than nine thousand species. Based on molecular and paleontological analysis, Alan Cooper and David Penny (1997) argued quite the opposite of Feduccia—not only that modern birds originated well before the K–Pg boundary (indicated by ~65 Ma on figure 7.11*B*), but that twenty-two lineages of modern birds crossed this boundary. The numbers given on this downward-opening phylogeny show various levels of confidence known as bootstrap values along each branch, where 100 is best. As these two phylogenies show, the issue of the timing of the origin of modern birds remains contentious. More recent studies (for example, Longrich, Tokaryk, and Field 2011) accept that fossils do indicate that neornithines existed in the Late Cretaceous, but the claim for a Late Cretaceous radiation of these modern birds as seen in Cooper and Penny cannot be supported by the fossil record (see figure 7.11*B*).

The issue of when modern mammals arose represents an even greater schism between paleontological and anatomical data, on one hand, and molecular data, on the other. The single greatest reason for this disagreement is that unlike more fragile

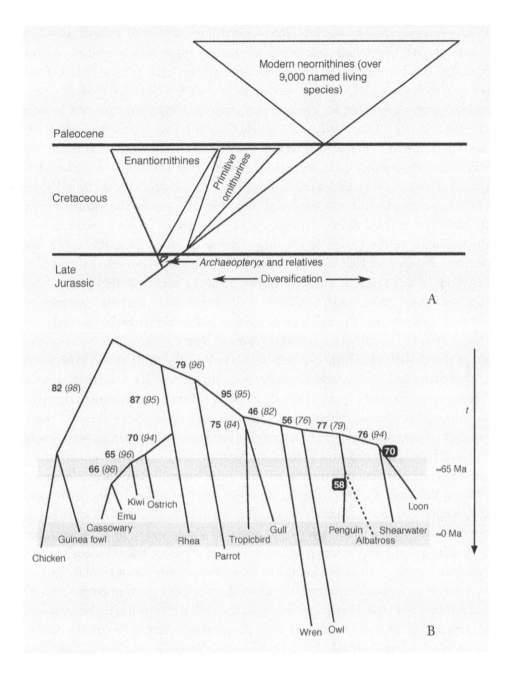

Modern neornithines (over
9,000 named living
species)

Paleocene

Enantiornithines

Primitive ornithurines

Cretaceous

Late
Jurassic

Archaeopteryx and relatives

←——— Diversification ———→

A

79 *(96)*

82 *(98)*

87 *(95)*

95 *(95)*

46 *(82)*

75 *(84)*

56 *(76)* 77 *(79)*

76 *(94)*

70 *(94)*

65 *(96)*

66 *(86)*

t

70

58

≈65 Ma

Kiwi Ostrich

Emu

Cassowary

Loon

≈0 Ma

Guinea fowl

Rhea

Gull

Penguin   Shearwater

Tropicbird

Albatross

Chicken

Parrot

Wren Owl

B

FIGURE 7.11 *(A)* Phylogeny, redrawn after one in Alan Feduccia's "Explosive Evolution in Tertiary Birds and Mammals" (1995) and *The Origin and Evolution of Birds* (1996), arguing that Neornithes did not originate and radiate until after the K–Pg boundary, and for a very substantial radiation and extinction of Enantiornithes, from J. David Archibald's *Extinction and Radiation* (2011); *(B)* Alan Cooper and David Penny's downward-opening phylogeny, arguing for many lineages of birds arising before the end of the Mesozoic, from "Mass Survival of Birds Across the Cretaceous–Tertiary Boundary" (1997). ([*A*] reproduced with permission of Johns Hopkins University Press; [*B*] reproduced with permission of the American Association for the Advancement of Science)

birds, for which the fossil record remains relatively sparse and hence less defensible as a source of data, mammals left a far more extensive fossil record because of their more durable bones and teeth. By the mid-nineteenth century, scholars identified tiny fossil jaws and teeth from the Mesozoic era that today we would agree represent mammals. Sorting out how these might be related to modern mammals remained more questionable. By the time of Charles Darwin in the latter half of the nineteenth century, as in the two mammalian phylogenies in his letter to Charles Lyell in 1860, the concept of marsupial and placental mammals began to emerge, although Darwin could not decide whether marsupials and placentals shared a nonmarsupial/nonplacental common ancestor (see figure 4.12) or whether primitive marsupials were the ancestors of both marsupials and placentals (see figure 4.13).

In his phylogeny, Ernst Haeckel (1866) seemed more certain of ancestry, showing marsupials (his Didelphia) as ancestors of modern marsupials as well as modern placentals (his Monodelphia), which arose somewhere in the Triassic period, if not before (see figure 5.4). Others speculated as well on the origin of modern mammalian groups; for example, Vladimir Kovalevskii's (1876) phylogeny of hooved mammals shows a hypothetical Urungulata from the Late Cretaceous, which we now know existed as *Protungulatum* (see figure 5.2C) (Archibald et al. 2011). By the mid-twentieth century, there seemed little doubt that marsupials and placentals possessed long, separate evolutionary histories. If we consider only placentals, their phylogeny began to take on a particular shape based largely on the fossil record of the time that enabled the modern groups of placentals to be traced back to only the beginning of the Cenozoic era, now dated at about 66 mya, although a few ancestral lineages trickled backward into the Cretaceous—an explosive radiation based on the fossil record. This iconography occurs in trees by Alfred Romer from 1933 (see figure 6.7B) and 1971 (see figure 6.7C) that characterized the basic ideas of the time: that much of the radiation of modern placental mammals occurred after the extinction of the dinosaurs, although more ancient eutherian ancestors or cousins of the modern placentals had been recognized in the fossil record of both the United States and Mongolia by the early twentieth century. The knowledge of such Cretaceous eutherians expanded rapidly in the last third of the twentieth century.

The advent of more sophisticated molecular techniques discussed earlier began rapidly to change the phylogenetic terrain. The vertebrate paleomammalogist Michael Novacek (1992) presented what at that time constituted the current state of affairs for eutherian and placental evolution in an aptly titled paper, "Mammalian Phylogeny: Shaking the Tree." In his phylogeny, he shows modern orders of placentals, with the exception of the problematic Insectivora, going back to only the beginning of the Cenozoic (figure 7.12A, *heavy lines*), much as Romer (1933, 1971) presents, but with quite a new wrinkle. Using lighter solid lines, Novacek traces various lineages backward into the Cretaceous, joining to form various super- or interordinal groupings, but, as he fully acknowledges, the actual dates of these splitting events could be inferred only from the relationships of the clades and their known ages. Nevertheless, his educated guess placed the earliest of the placental splits at about 115 mya (indicated by arrow and *p*) and pushed back the origin of eutherians a few more million years to slightly over 120 mya (indicated by arrow and *e*). The abstract for the article clearly conveys the intended message and the then state of phylogenetic affairs: "Recent palaeontological discoveries and the correspondence between molecular and morphological results provide fresh insight on the deep structure of mammalian phylogeny. This new wave of research, however, has yet to resolve some important issues" (121). As Novacek relates in the article and shows on his phylogeny, as of 1992 various molecularly and morphologically based studies converged on the following ideas:

1. New World edentates (armadillos, sloths, anteaters) perhaps represent the earliest branching clade among placentals.
2. The basic relationships of the artiodactyls (cows, deer, sheep, pigs, camels) can be ascertained.
3. Artiodactyls and cetaceans cluster together separate of several other placental orders.

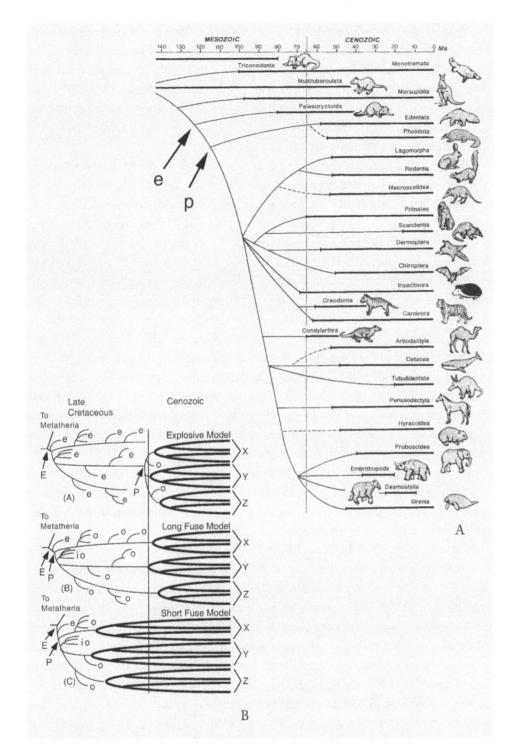

A

B

FIGURE 7.12 (*A*) Michael Novacek's phylogeny showing eutherians
(*e* [added]) and placentals
(*p* [added]), in which modern orders
of placentals go back to only the
beginning of the Cenozoic (*heavy
lines*), while some lineages date to
the Cretaceous (*light lines*), joining
to form various super- or interordinal groupings, from "Mammalian
Phylogeny" (1992); (*B*) J. David
Archibald and Douglas Deutschman's
names for three patterns of eutherian and placental evolution that
had been recognized (*E* = origin of
Eutheria; *P* = origin of Placentalia;
e = eutherian but not placental
lineage; *o* = extant placental order;
io = interordinal placental groups),
from "Quantitative Analysis of the
Timing of Origin of Extant Placental
Orders" (2001). ([*A*] reproduced
with permission of the Nature Publishing Group; [*B*] reproduced with
permission of Springer)

A clearer understanding of how the various modern orders of placental mammals related one to another and an idea of when these splittings might have
occurred awaited newer molecular sequencing techniques. Already, however, the
sort of trees that Romer showed, on one hand, and Novacek represented, on
the other, portrayed two views of the timing of placental origination and diversification. In 2001, Douglas Deutschman and I named these two different views

the Explosive and the Long Fuse models. The Explosive model, as in Romer's trees, shows that placental mammals originated very close to either side of the K–Pg boundary (*P* in figure 7.12*B*), with modern orders appearing very soon after (*o* in figure 7.12*B*). Contrast this with the Long Fuse model, as in Novacek's tree, which shows that placental mammals originated quite far back in the Late Cretaceous (*P* in figure 7.12*B*), but as in the Explosive model, modern orders appear near the K–Pg boundary (*o* in figure 7.12*B*).

A series of papers in the early twenty-first century offered new and sometimes startling phylogenetic results. William Murphy and colleagues (2001) used molecular data to argue that modern orders of placental mammals fall into four larger, superordinal groupings. Three of these groupings offered no great surprises, as they accorded in general with paleontological data (figure 7.13*A*), but the fourth, Afrotheria, was a surprise. It united typically African mammals, from aardvarks to elephants to golden moles, into one group. The analysis and phylogeny show this radiation of largely African mammals (see figure 7.13*B*), which had not been predicted on paleontological and morphological data, although sirenians, elephants, and hyraxes previously clustered with such data. Further, in the biogeographic tradition of Haeckel in the nineteenth century and Brundin in the twentieth, Murphy and his colleagues overlaid this superordinal phylogeny onto what was understood of the breaking apart of the southern supercontinent Gondwana to suggest how the phylogeny came to be (see figure 7.13*C*). They argued that placentals arose and began to divide into four superorders between 108 and 101 mya, in keeping with the splitting of Africa and South America between 120 and 100 mya. Certainly, then, their phylogenetic analysis argued for at least the Long Fuse model, according to which the placentals originated and began to diversify at least into superordinal groups in the Late Cretaceous (see figure 7.12*A*), but when modern orders of placentals arose remained an open question.

An article by Olaf Bininda-Emonds and colleagues (2007) used molecular data to present a supertree, meaning that it was constructed by clustering a series of smaller trees using various data sets. According to these authors, "The supertree contains 4,510 of the 4,554 extant species recorded . . . making it 99.0% complete at the species level" (507). They argued that not only did placentals and other modern mammals originate and split into superorders in the Late Cretaceous, but modern orders of mammals also appeared in the Late Cretaceous, well before the extinction of the dinosaurs some 66 mya. The authors not only dated the origination and earliest diversification of mammalian orders to some 93 mya, but also argued that mass extinction at the end of the Cretaceous did not have a major, direct effect on the diversification of modern mammals. This follows the pattern of the Short Fuse model of the origination and early diversification of all modern mammals within the Late Cretaceous (see figure 7.12*B*). The spiraling rather than tree-like phylogeny used by Bininda-Emonds and his colleagues, while not new or even a rare method for representation, certainly was chosen in part because of the need to show a considerable number of species (figure 7.14). The dashed circular line on the phylogeny marks the K–Pg boundary at about 66 mya. This means that not just the ancient ancestors of some families of living mammals but actual members of these families lived with dinosaurs—rats, mice, and beavers cavorted at their feet

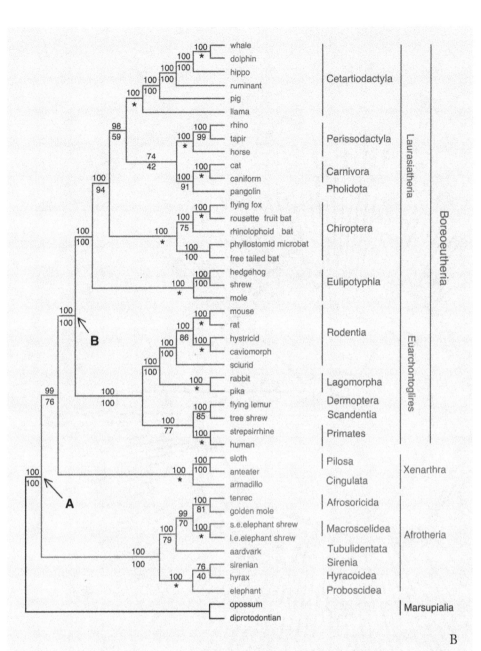

B

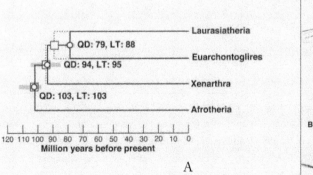

A

C

FIGURE 7.13 Using molecular data, William Murphy and his colleagues found (A) four major superordinal clades of (B) living placental mammals, which they argued (C) show a split following patterns of the breakup of Gondwana, from "Resolution of the Early Placental Mammal Radiation Using Bayesian Phylogenetics" (2001). (Reproduced with permission of the American Association for the Advancement of Science)

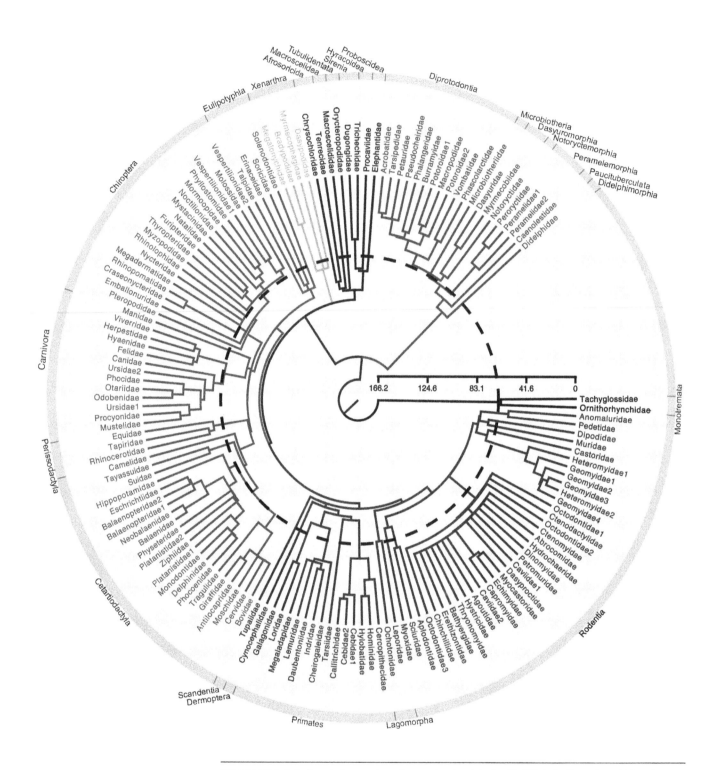

FIGURE 7.14 Spiraling phylogeny, based on molecular data, by Olaf Bininda-Emonds and his colleagues, arguing that not only did modern mammals originate and divide into superorders in the Late Cretaceous, but modern orders of mammals also appeared in the Late Cretaceous, well before the extinction of the dinosaurs (*dashed circle*), from "The Delayed Rise of Present-Day Mammals" (2007). (Reproduced with permission of the Nature Publishing Group)

while camels shared their fields, and varieties of bats flew above them—a scenario that many biologists reject.

The pendulum swung fully with a paper by Maureen O'Leary and colleagues (2013a). The authors combined morphological, or what they termed phenomic, characters in obvious apposition to genomic characters, along with molecular data. While the results shown in their phylogenetic tree did differ in some ways from those in earlier studies, the biggest furor resulted from marked differences in the timing of these evolutionary events (figure 7.15). Whereas Murphy and colleagues (2001) had placed at least the origin of placental superorders at about 100 mya and Bininda-Emonds and co-workers (2007) had further argued that modern orders of mammals appeared by 93 mya, O'Leary and her colleagues not only dated the origin of modern placental orders to after the K–Pg boundary, at some 66 mya, but also argued that placental mammals did not even appear until then. Thus, according to them, placentals arose some 36 million years later than Bininda-Emonds and his colleagues determined, not an insignificant interval, given that all researchers would agree that the remaining time span for placentals on Earth is some 66 million years in duration. In the O'Leary scenario, unlike that of Bininda-Emonds, the end of the Cretaceous had a profound effect on mammalian evolution, and contrary to that of Murphy, the breaking apart of Gondwana was not a factor in early placental diversification.

Any differences in their phylogenetic results and trees pale in comparison with the profoundly contrasting views of the timing of these events. For these varying interpretations, this difference of some 36 million years is far from insignificant, and much of it derives from divergent views about the importance of assuming or not assuming some sort of molecular clock (O'Leary et al. 2013b; Springer et al. 2013) as well as the way fossils can be used to calibrate age estimates for the various lineages. Mario dos Reis, Philip C. J. Donoghue, and Ziheng Yang (2014) presented a series of four trees showing identical evolutionary relationships among but greatly different timings for the origin of modern placental mammals (figure 7.16), echoing the older origination espoused by Bininda-Emonds and colleagues (2007) versus the more recent origination argued by O'Leary and co-workers (2013a). In figure 7.16, note the position of the line labeled "Placentalia," which varies from almost 100 mya to about 65 mya in the placement of the time of origin for this group. Regardless of whose interpretation of the timing of the origination of modern placental mammals is correct, we see an important use of phylogenetic trees (or cladograms) to present differing ideas about origination times. This recalls Haeckel's (1866) use of trees to offer various hypotheses about the origin of life (see figure 5.3). Once again, we see trees showing something beyond evolutionary relationships.

These ongoing disputes about the timing of major evolutionary events occur for all major groups, not just for the study of mammalian evolution and diversification. These considerable differences in opinion concerning the timing of events using various phylogenetic analyses, whether they deal with the origin of multicellular life or the origin and diversification of modern birds and mammals, will assuredly continue. Notwithstanding, the question of timing is not the only issue dogging tree building; the more profound question arises as to whether a tree is even a reasonable visual metaphor for the history of life.

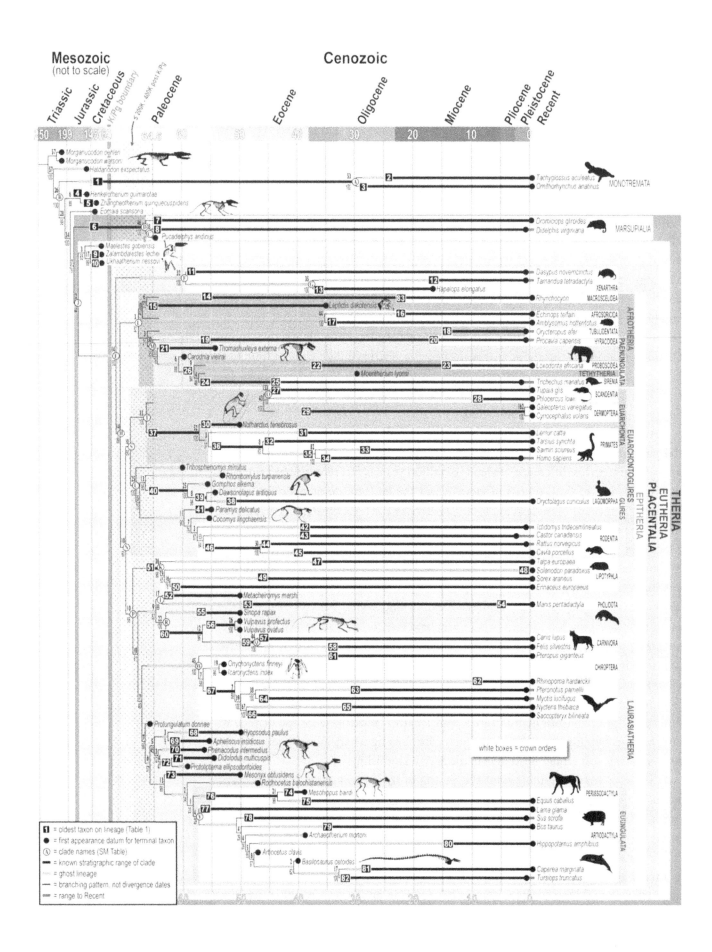

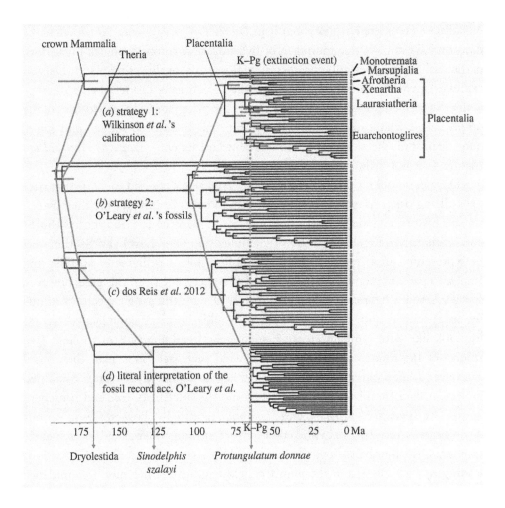

crown Mammalia   Placentalia

Theria   K–Pg (extinction event)   Monotremata
Marsupialia
Afrotheria
Xenartha
Laurasiatheria   Placentalia
(a) strategy 1:
Wilkinson *et al.*'s
calibration   Euarchontoglires

(b) strategy 2:
O'Leary *et al.*'s fossils

(c) dos Reis *et al.* 2012

(d) literal interpretation of the
fossil record acc. O'Leary *et al.*

175   150   125   100   75 K–Pg 50   25   0 Ma

Dryolestida   *Sinodelphis*   *Protungulatum donnae*
szalayi

FIGURE 7.16 Four phylogenetic trees by Mario dos Reis, Philip C. J. Donoghue, and Ziheng Yang, showing the same evolutionary relationships among but very different times for the origin of Placentalia, ranging from almost 100 million years ago to about 65 million years ago, from "Neither Phylogenomic nor Palaeontological Data Support a Palaeogene Origin of Placental Mammals," *Biology Letters* 10 (2014):fig. 1. (Reproduced with permission of the Royal Society)

The Molecular Tree Becomes Tangled

The advent of molecular systematics showed that at least for simpler organisms, such as bacteria, the tree became a tangled, thorny bramble menacing results everywhere: "It is interesting to contemplate a tangled bank, clothed with many plants of many kinds" (Darwin 1859:489). When Darwin penned these words in the final sentence of *On the Origin of Species*, he was reflecting on complex interaction of species in ecological relationships that produced evolutionary change over millions of generations. He could not have realized just how tangled the evolutionary history of many species could be, enough in fact for some biologists to cast serious doubt on the tree metaphor for the history of life. Darwin certainly knew of and wrote about at least one form of evolutionary entanglement, hybridization,

FIGURE 7.15 Phylogenetic tree by Maureen O'Leary and her colleagues, based on combined phenomic and genomic characters, differs in some ways from earlier diagrams, primarily in the placement of the origin of placentals, including all extant orders, after the K–Pg boundary some 66 million years ago, from "The Placental Mammal Ancestor and the Post-K-Pg Radiation of Placentals" (2013). (Reproduced with permission of the American Association for the Advancement of Science)

including a complete chapter about it in *On the Origin of Species*. Nonetheless, he did not fully know the importance of hybridization among complex organisms, particularly in the speciation process of angiosperms, or flowering plants. Darwin expressed great consternation concerning the evolutionary history of angiosperms, which he termed an "abominable mystery." He struggled with the idea that the origin and diversification of flowering plants might be both sudden and rapid, quite against his view that evolution was slow and stately—*natura non facit saltum* (nature does not make a leap) (Friedman 2009).

Darwin would be thrilled and possibly even not surprised to learn that perhaps all angiosperms experienced some form of hybridization during their evolutionary history, and angiosperms may number as many as 400,000 species (Soltis and Soltis 2009). Figure 7.17*A* shows a simple example of how hybridization may be represented in a tree, somewhat resembling how a pedigree of an offspring from two parents might look (Linder and Rieseberg 2004). In this case, however, species *X* and *Y* hybridize to produce species *B*, with the parent species continuing as species *A* and *C*. But what, then, would the more complicated tree of life look like with rampant crossing, such as between angiosperm species? Two studies of the numerous examples show what such trees look like. The annual sunflower *Helianthus* appears in such a tree (see figure 7.17*B*), using combined molecular data from both the chloroplast and the nucleus (Gross and Rieseberg 2005). Two species, *H. petiolaris* and *H. annuus*, repeatedly show crosses in various environments, resulting in hybrids that fare better than either parent in such environments. The numbers above the horizontal lines show the number of mutations leading to each clade, and the numbers below indicate a measure of confidence in that particular clade, in this case using a technique known as bootstrapping, in which 100 is the maximum value. Once one knows what is being shown, the hybridization events in this tree become quite clear. The second, even more complex example involves the small annual plant *Clarkia*, named for the explorer William Clark of Lewis and Clark fame and found mostly in the western United States (Futuyma 1986). Figure 7.17*C* shows the intricate phylogeny and hybridization within this genus. For the parent species, the numbers refer to the number of chromosomes in the gametes, which are not always the same in both parent species, in turn giving rise to the number of chromosomes in the descendant hybrid (for example, *C. mildrediae*-7 and *C. virgata*-5 yielded *C. rhomboidea*-12). Interestingly, the sort of analysis shown in this phylogeny allows a prediction of what the gamete chromosomal number would be for unknown ancestors, indicated by numbers unaccompanied by names. Although seemingly a tangled network, this powerful phylogenetic hypothesis uses a tree to argue in which direction and by which means (hybridization in this case) evolution has traversed.

Because the very basis of inheritance was unknown to Darwin and most others in the nineteenth century, they could not have conceived how important other methods of genetic transfer were between organisms of the same and different species. And anyway, how would one show this on the traditional metaphorical tree of life? Although various processes had been known for some time, dubbed "reticulate evolution" for obvious reasons, the issue came to a head in 1999 when the biochemist W. Ford Doolittle (b. 1942) published two of what he termed "crude

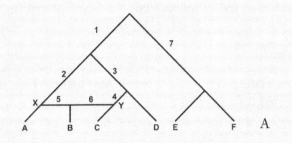

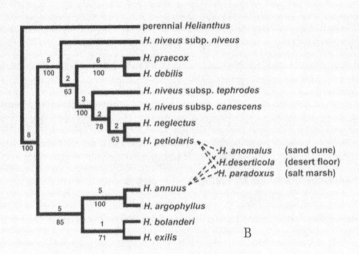

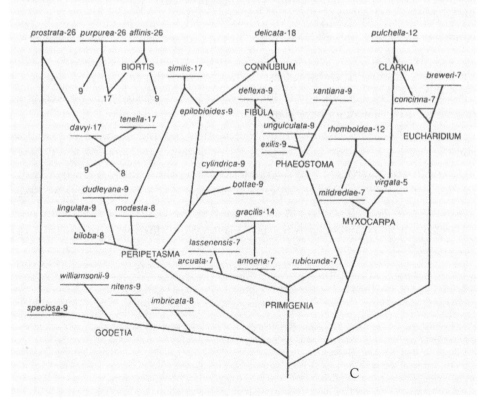

FIGURE 7.17 (A) C. Randal Linder and Loren Rieseberg's simple example of how hybridization may be represented in a tree, from "Reconstructing Patterns of Reticulate Evolution in Plants" (2004); (B) Brianna Gross and Rieseberg's phylogenetic analysis of the annual sunflower *Helianthus*, using combined molecular data from both the chloroplast and the nucleus, showing repeated hybridization from *H. petiolaris* and *H. annuus*, from "The Ecological Genetics of Homoploid Hybrid Speciation" (2005); (C) Douglas Futuyma's more complex example, showing the intricate phylogeny and hybridization within the genus *Clarkia*, from *Evolutionary Biology* (1986). ([A] reproduced with permission of the Botanical Society of America; [B] reproduced with permission of the *Journal of Heredity* and L. Rieseberg; [C] reproduced with permission of Sinauer)

sketches[s]." Crude they were, but powerful as well in their simple message—much reticulate evolution occurred in the history of life, especially the early part, in the form of horizontal or lateral gene transfer. This greatly complicates our attempts to reconstruct the history of life on Earth. If not a tree, what metaphor should be used?

Doolittle's (1999) first sketch represented what at the time was "commonly taken as a representation of organismal phylogeny and the basis for a natural classification" (2125). At the top of this sketch, we see the three so-called domains to which all but viruses were believed to belong (figure 7.18*A*). In the diagram, Bacteria encompass what we traditionally consider bacteria to be, such as *Escherichia coli*, commonly called *E. coli*, which strikes fear into the hearts, or guts, of many people, even though many strains are harmless or even beneficial. In this classification, Bacteria also include a great variety of other prokaryotic microorganisms, such as blue-green algae, all of which lack a nucleus (Gk. *karyon*) or other cellular organelles, such as mitochondria. Archaea, once thought to be a form of bacteria and far less known to most people, are also prokaryotic microorganisms, many of which live in extreme conditions—the original thrill seekers. The third domain, Eukarya or Eukaryota, whose cells possess a nucleus and cellular organelles, includes animals, plants, and fungi, as shown in Doolittle's sketch. Note, however, that two branches of Bacteria trace to Eukarya. Doolittle did not intend to indicate, as a nineteenth-century biologist such as Haeckel might have meant, that these lineages had separately evolved into eukaryotes but something more radical, first considered at the beginning of the twentieth century but not validated until the 1960s. In 1967, the biologist Lynn Margolis (1938–2011) provided microbiological evidence that early in evolutionary history, eukaryotic cells had incorporated various bacteria into themselves—mitochondria in animals and plants, and chloroplasts in plants—which became known as the widely accepted endosymbiotic theory (Sagan 1967). Thus by at least the mid-twentieth century, there was some inkling that the tree as a metaphor for the history of life existed on borrowed time, at least for part of evolutionary history.

It was Doolittle's second "crude sketch" that brought to the fore just how complicated the history of life on planet Earth really is (see figure 7.18*B*). Granted, much of this entanglement may have begun early in the history of life, but "early" in the frame of Earth's history represents perhaps 4 billion years, or only three-quarters of the time that life has existed on this planet, compared with the possibly 1 billion years that eukaryotes have been around. Instead of just two arrows

FIGURE 7.18 W. Ford Doolittle's sketches represented (*A*) what was considered the common view of organismal phylogeny, compared with (*B*) what was being revealed about horizontal gene transfer between many simpler species, from "Phylogenetic Classification and Universal Tree" (1999); (*C*) Carl Woese and George Fox's three domains, before the name change from Archaebacteria to Archaea, from Woese, "Bacterial Evolution" (1987); (*D*) the navigational tree-like figure from the Tree of Life (ToL) Web Project. ([*A*] and [*B*] reproduced with permission of the American Association for the Advancement of Science; [*C*] reproduced with permission of *Microbiology and Molecular Biology Reviews*; [*D*] reproduced with the permission of Nick Kurzenko)

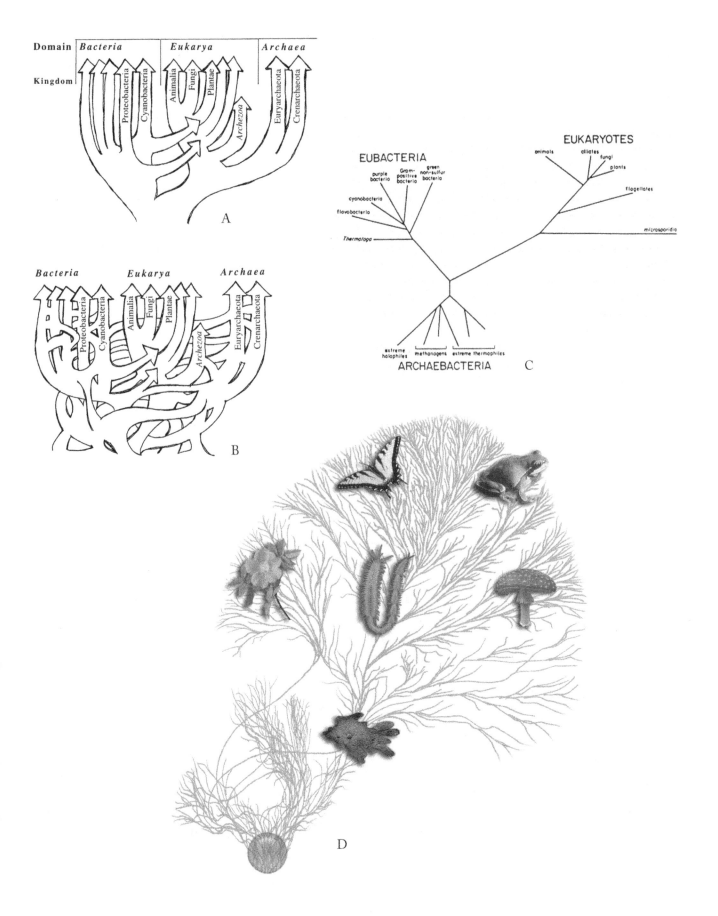

A

B

Domain | *Bacteria* | *Eukarya* | *Archaea*

Kingdom

Proteobacteria
Cyanobacteria
Animalia
Fungi
Plantae
Archezoa
Euryarchaeota
Crenarchaeota

Bacteria   *Eukarya*   *Archaea*

Proteobacteria
Cyanobacteria
Animalia
Fungi
Plantae
Archezoa
Euryarchaeota
Crenarchaeota

EUBACTERIA

purple
bacteria
Gram-
positive
bacteria
green
non-sulfur
bacteria
cyanobacteria
flavobacteria
Thermotoga

EUKARYOTES

animals
ciliates
fungi
plants
flagellates
microsporidia

extreme
halophiles
methanogens
extreme thermophiles

ARCHAEBACTERIA

C

D

heading from bacteria to eukaryotes, the tree now resembles nothing less than a tangled mess of interdomain relationships. It became clear that during the history of life, the basal members of these domains began to sort themselves out but along the way also cross-shared considerable chunks of their genomic and cellular machinery. The entanglement of early life means that the interrelationships of the three domains, and whether the contents of each domain are correctly assigned, remain an unsolved puzzle.

The three domains were not even recognized until 1977, when Carl Woese (1928–2012) and his colleague George E. Fox (b. 1945) distinguished a new group based on 16S ribosomal RNA (the molecular scaffold in Archaea and Bacteria used for synthesizing proteins), which they first called Archaebacteria but later changed to simply Archaea to make it clear that this new domain was not a form of bacteria. Woese (1987) presented the major groups as an unrooted network in a review paper in which Archaea was still referred to as Archaebacteria (see figure 7.18C), rather than a tangled tree as in Doolittle's (1999) figure (see figure 7.18B). Even the rooting of the tree cannot be done with any certainty because an outside nonliving comparison is needed—a rock, for instance. Aristotle (1991, 2007) and Charles Bonnet (1745) had it more right than wrong in placing inanimate material at the base of all life or, more correctly, as the sister group to life.

The most current treatments of how the relationships of life on Earth should be represented grapple with such questions as rooting the tree of life and the complexities of tree reconstructions caused by horizontal gene transfer. A special issue of the journal *Biology & Philosophy* titled "The Tree of Life" examined "scientific, historical and philosophical aspects of debates about the Tree of Life, with the aim of turning these criticisms towards a reconstruction of prokaryote phylogeny and even some aspects of the standard evolutionary understanding of eukaryotes" (O'Malley, Martin, and Dupré 2010:441). Although the contributors varied considerably in their views about the metaphorical relevance of a tree of life image, the journal certainly placed the issues front and center.

In the rapidly changing world of the Internet, a number of sites examine the tree of life, with the Tree of Life (ToL) Web Project (Maddison and Schulz 2007; Maddison, Schulz, and Maddison 2007), an unarguably major one. Its home page is graced by a visually poetic image, which the authors emphasize is a greatly simplified illustration that serves a navigational purpose (the tree is a table of contents of sorts) (see figure 7.18D). The authors stress that the earliest history of life is obscure in the diagram because evidence indicates that rather than a single primordial cell, the first organisms possessed loosely constructed cellular organization and frequently exchanged genetic material; perhaps the origin of life cannot be seen as monophyletic or polyphyletic but as a prebiotic amalgam exchanging information. Even later, after the establishment of distinct genetic lineages, these exchanges of genetic material continued, which the diagram indicates by lines connecting distant branches of the tree. Whereas horizontal gene transfer clearly helped define Eubacteria and Archaea, its significance for eukaryotic evolution becomes less clear with the exception of eukaryotes' acquisition of organelles such as mitochondria and chloroplasts by the engulfment (endosymbiosis) of bacteria. As shown in figure 7.18D, the relative magnitudes of various clades distort an accurate representation of diversification; if not, arthropods (notably insects) would greatly dominate the figure. Based on recent estimates of the true diversity of bacteria, protists, fungi, and nematodes, the remainder of the tree would be filled by branches for these groups; vertebrates and plants would be barely visible to the naked eye. Once again, the importance of humans is placed in its correct perspective.

Coming Full Circle

Near the end of the twentieth century, with a combination of new and faster methods to extract and sequence molecular data combined with ever accelerating computer power, the phylogenetic trees began to take on a life of their own—no longer simply bifurcating diagrams. Additionally, with the simple yet profound realization that relative splitting on a tree, and not nearness of groups or how high or low on a tree they rest, determines evolutionary relationship, a new freedom in how to express trees arose. I use the word "tree" broadly because many of the new diagrams are anything but trees in a traditional sense. As seen, they vary from trees to circles to webs. All of them are the same in using relative splitting of groups or clades to show evolutionary relationships, but some use anatomy, others use molecular data, and others use both. In this way, these figures are the same. In part, some of the new shapes arose out of necessity to conserve space yet present as much information as possible.

The traditional tree form, even if festooned with fossil species, presents a basic triangular shape that wastes space on the page, but less so with a circle or, more accurately, a spire on which the relationships spiral inwardly to the origin with the various taxa arrayed around the circumference. A well-known version of this sort of figure appeared in *Science* (Pennisi 2003), by the molecular systematist David Hillis (b. 1958) and colleagues (figure 7.19*A*). This densely packed figure shows three thousand named species throughout all the kingdoms of extant life and is best viewed when scaled to a diameter of 5 feet (1.5 m). Hillis writes, "The number of species represented is approximately the square-root of the number of species thought to exist on Earth (i.e., three thousand out of an estimated nine million species), or about 0.18% of the 1.7 million species that have been formally described and named" (Hillis, Zwickl, and Gutell). Relative proportions of various groups skew toward eukaryotes, including us. Even then, it is humbling for *Homo sapiens* to be one infinitesimal smudge labeled "You are here." It is reminiscent of the famous photograph taken from *Voyager 1* near the edge of the solar system showing Earth as a small pinprick of light.

Although intriguing in their form, these circular representations of life's diversity are not new. A favorite is the frontispiece of Heinrich Bronn's (1850–1856) atlas for the third edition of his book *Lethaea geognostica* (see figure 7.19*B*). In the original edition, Bronn (1837–1838) used the more traditional paleontological chart showing older fossils at the bottom and younger ones near the top (see figure 3.13*B*), but in this over 150-year-old version he attempted, as do systematists of today, to save space by using a circular form, with the oldest ages of Earth near the center and life-forms radiating outward to the circumference. This is by no

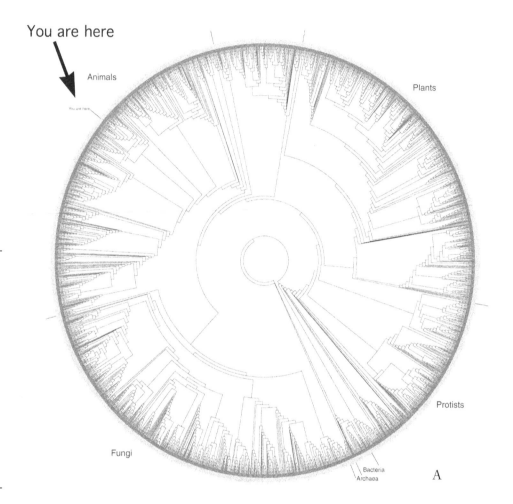

You are here

Animals

Plants

Protists

Fungi

Bacteria
Archaea

A

FIGURE 7.19 (A) Spiral phylogeny by David Hillis, Derrick Zwickl, and Robin Gutell, showing three thousand named species throughout all kingdoms of extant life; (B) frontispiece of Heinrich Bronn's atlas for *Lethaea geognostica* (1850–1856), a circular paleontological range chart (not phylogeny) of life's diversity. ([A] David Hillis, Derrick Zwickl, and Robin Gutell, University of Texas, are thanked for allowing the use of the spiral phylogeny)

means a phylogeny but a circular paleontological range chart, but recall that in 1858 Bronn published only the second tree-like figure, clearly advocating evolutionary descent; Jean-Baptiste Lamarck's (1809) version was the first. One can imagine that if Bronn had not died in 1862, we might have seen more evolutionarily based paleontological diagrams from him. Near the outer edge of the circle, Bronn indicates the number of extant species for that particular group, with a total of more than 173,000, a far cry from the millions of species we now believe

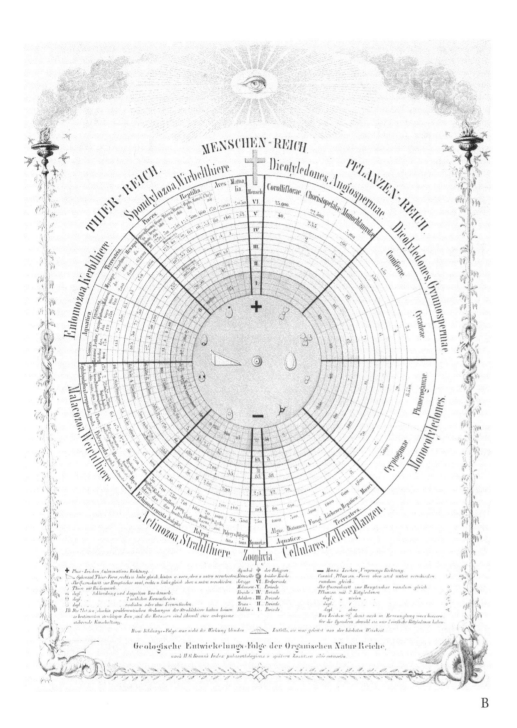

B

to exist, but nevertheless impressive. Of course, unlike the diagram of Hillis and his colleagues (see figure 7.19*A*), humans (*Menschen*) rest at the top of Bronn's circle, crowned by a crucifix and overseen by the eye of God. No matter how one chooses to portray our position in nature, we are part of it. Our interpretations of this position, at times messy and sometimes downright nasty, help us divine who we think we are and where we believe we belong within the scheme of things. It is an ongoing process.

CHAPTER EIGHT

The Paragon of Animals

Where do we or, more broadly, where does our species, *Homo sapiens*, belong in this nexus of relationships? Are we the pinnacle of creation espoused by the ancients, and not so ancients, or do we fit as one small twig on the vast tree (or web) of evolutionary relationships? In large measure, one's worldview dictates the answer; but in spite of what we as individuals believe about this vexing issue, our understanding of our place in nature has expanded in the past two thousand years almost beyond comprehension, to the point that we now realize that we constitute but a minuscule part of life on Earth. Some deny this, some revel in it, but most do not even think about it. To a great extent, the perceptions of our place in nature evolved over time as we discovered and attempted to understand the vast living and nonliving world that surrounds and defines us.

What metaphor best describes our place in the profusion of animate nature? It depends. Are we these unique creatures cast in God's image, just below the angels? We recall Hamlet's musing in act 2, scene 2, of Shakespeare's *Hamlet* (or similarly in the rock-opera *Hair*): "What a piece of work is a man! How noble in reason, how infinite in faculty! In form and moving how express and admirable! In action how like an Angel! In apprehension how like a god! The beauty of the world! The paragon of animals!" This is certainly how the vast majority of people view humans' place in nature, even when accepting that evolution occurred. Not much has changed in our overall perceptions of ourselves stretching back to the Aristotelian *scala naturae*, although as figured, this concept of humans in a biological context dates from only the eighteenth century with Charles Bonnet's (1745, 1781) representations of humans at the top of the ladder (see figure 1.4) or the stairway (see figure 1.5). Even the first attempts at placing humans in a tree figure were of a nonevolutionary sort, such as those by Carl Edward von Eichwald in 1829 (see figure 3.3) and Edward Hitchcock in 1840 (see figure 3.11).

Our philosopher Hamlet's stirring words may treat humanity as the "paragon of animals," but it does not end there. In fact, this praise is a sham: "And yet to me, what is this quintessence of dust? Man delights not me, no, nor woman either." Although Shakespeare's intent lies elsewhere for poor mad Hamlet, it does serve the warning that biologically humans really are not the pinnacle of evolution whether at the top of the ladder or perched in the tree, except possibly in our minds and in our definitions. We are but a twig on the bush of human evolutionary history that becomes less and less certain with the discovery of each new fossil hominid and the completion of each new genetic sequence. For many this is troubling, but for many others it is liberating.

Grades, Clades, and Races

Humans, as do most other species, attempt to differentiate "us from them." Some birds use song to distinguish between neighbors, dogs may sniff their canine encounters, but humans as visually oriented mammals largely use sight to tell differences among humans. No wonder, then, that humans from their very beginnings visually recognized both physical and cultural distinctions among one another. One of the earliest tree-like attempts at visualizing supposed racial differences as well as the ancestry of all living humans dates from at least the early seventeenth century, often as an engraving in early copies of the King James Bible. This representation is biblical, not scientific, showing in an unmistakable tree form the sons of Noah—Japheth, Shem, and Ham (Iaphet, Sem, and Cham in figure 8.1*A*)— giving rise to peoples of Europe, Asia, and Africa, respectively. In this scene, the ark rests on Mount Ararat, Armenia (present-day Turkey), with the tree springing from Noah and his ark. The "Hebrewes" at the top are likely shown as the Chosen People of the Hebrew Bible. Although a proverbial kernel of truth might be found in such a tree, it purports to show the ancestry of all humankind coming from Noah.

Such a view of human history within the Western tradition lasted well into the nineteenth century, but among many educated people it began to crumble by time of the publication of *On the Origin of Species*. Critics quickly lambasted Darwin for toppling humans from the pinnacle of creation. The only problem was that Darwin (1859) never mentioned human origins in this book, other than his cryptic sentence on the third to last page: "Light will be thrown on the origin of man and his history" (488). Not until *The Descent of Man, and Selection in Relation to Sex* (1871) did he broach the topic when he presciently predicted, "It is therefore probable that Africa was formerly inhabited by extinct apes closely allied to the gorilla and chimpanzee; and as these two species are now man's nearest allies, it is somewhat more probable that our early progenitors lived on the African continent than elsewhere" (199).

The only known evolutionary trees including humans that Darwin produced were his unpublished sketchs (see figures 4.15 and 4.16), the version in 1868 possibly from the influence of St. George Mivart's (1865, 1867) published primate trees (see figure 5.1). Nevertheless, Darwin's (1859) volume opened the floodgates. Only six

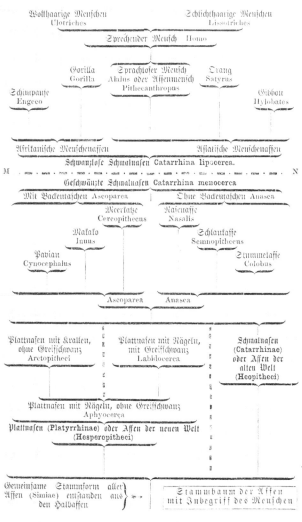

A

B

FIGURE 8.1 (A) Noah and his descendants (ca. 1611); (B) Ernst Haeckel's "Family Tree of Monkeys with Humans," from *Natürliche Schöpfungsgeschichte* (1868).

months later, Thomas Henry Huxley and Samuel Wilberforce proved to be the most vociferous participants in a heated discussion in Oxford when Wilberforce purportedly asked whether Huxley's ancestry from apes came through his mother's or his father's side, and Huxley retorted something to the effect that he would rather claim an ape as an ancestor than someone such as Wilberforce who abused his considerable debating powers on such important matters. Whatever the truth of the actual exchange, a number of Huxley's published essays, especially in 1861 and 1862, spelled out his views on human ancestry.

In 1863, Huxley's slim, 159-page *Evidence as to Man's Place in Nature* appeared. It included thirty-two figures, possibly the best known being the frontispiece (see figure 1.6*A*). Not calling it a progression, he later writes of species lying below humans on a scale of evolution. He does not present any phylogenetic trees. Although a considerable amount of the book's content appeared in earlier essays, *Evidence* reached an even wider audience. In the book, Huxley clearly aligns humans with apes. He explains that through embryological development, humans, as with all animals, start life as a fertilized egg that proceeds through various stages of vertebrate development and ends in a form far closer to apes than to any other animal. Also in 1863, Darwin's geological mentor and friend Charles Lyell published *Geological Evidences of the Antiquity of Man*, which examined early humans. Unlike Huxley's book, which discussed humans within a biological context, Lyell's dealt with early humans in Europe in more of an archaeological context. Although Lyell had come around to accepting some aspects of Darwin's descent with modification by means of natural selection, he remained, as Janet Browne (2002) notes, a "reluctant evolutionist."

The younger Darwinian adherents St. George Mivart and Ernst Haeckel first attempted to place humans in the context of evolutionary trees. The earliest such human-inhabited trees belong to Mivart (1865, 1867), who includes only living primates based on the axial and appendicular skeletons, respectively (see figure 5.1), and to Haeckel (1866), whose mammal tree includes humans and the fossil ape *Dryopithecus* (see figure 5.4). Haeckel unmistakably places *Dryopithecus* on the human lineage in his tree, yet in his classification it belongs to the same family as living great apes. This ambiguity becomes somewhat clearer in his text when he notes that although it can be classified with other living apes, "the recently discovered *Dryopithecus* fossil represents a very important form between the gorilla and man" (2:155). Haeckel hedges his bets here. He continues this theme on the next page, again noting that whereas *Dryopithecus* may be classified as an ape, it cannot be excluded from human ancestry.

Dryopitheus presents an interesting history. Éduoard Lartet (1801–1871) named and described it in 1856 as a fossil ape, but not a human relative, based on a partial limb bone, some teeth, and a lower jaw from a site in southwestern France. Today, various species of *Dryopithecus* occur at sites in Europe, Africa, and Asia, possibly dating from 12 to 9 mya, and although placed among apes it begs the question of if or how it may relate to human ancestry, given that chimps and humans share an ancestor perhaps between 13 and 7 mya based on current genomic and fossil data (Begun 2009).

Haeckel (1866) includes in his classification of the human family (Erecta) the genus *Homo* along with *Pithecanthropus*, a then undiscovered, hypothetical ancestor. To my knowledge, this represents Haeckel's first mention of this genus, at least in a book. In *Natürliche Schöpfungsgeschichte* (*The History of Creation*, 1868:514), Haeckel hypothesized species names such as *Pithecanthropus primigenus*, or *Homo primigenus*, and the better-known *Pithecanthropus alalus*, which he includes in his very stylized "Family Tree of Monkeys with Humans" (see figure 8.1*B*). He further hypothesized that fossils of these forms might be discovered in what is now Indonesia. Amazingly, following Haeckel's predictions, in 1891 the Dutchman Eugène Dubois (1858–1940) recovered fossil remains in Java that he named *Pithecanthropus erectus*, or Java Man, now known as *Homo erectus*. In addition to placing *Pithecanthropus* as ancestral to *Homo*, Haeckel divides "speaking man" (*Homo*) into "wooly-haired" (Ulotriches) and "straight-haired" (Lissotriches) modern humans. Late in the same book, Haeckel further divides wooly-haired and straight-haired humans into ten species and twenty races. In the volume's final figure, Haeckel presents a family tree of his human species and races. He does not stop here, for in the second edition of *Natürliche Schöpfungsgeschichte*

(1870) he further divides humans into twelve species and thirty-six races. The tree shown in figure 8.2*A* comes from E. Ray Lankester's translation (Haeckel 1876). Recall that Haeckel (1870) presented his map of human migration using a modified tree-like form originating in his hypothetical Lemuria in the Indian Ocean (see figure 5.7).

Cast in today's sensibilities and understanding of species concepts, Haeckel's ideas might be offensive as well as absurd. Within his era, however, Haeckel's views on human races varied only slightly from the societal norms for both evolutionists and nonevolutionists. In addition, Haeckel has been portrayed as a harbinger or even an enabler of Nazi racial propaganda, a charge that Robert Richards (2008) ably rebuts in his biography of Haeckel. As to the notion of multiple species of humans, even in Haeckel's day this was not a widely held view. Part of the problem, which still plagues us, concerns the more general issue of what constitutes a species. For example, throughout *On the Origin of Species*, Darwin struggled with differentiating between varieties and species. Given this, the importance of Haeckel derives from his being one of the first scientists to attempt to trace human evolution, even

FIGURE 8.2 Haeckel's (*A*) "Pedigree of the Twelve Species of Men," from *The History of Creation* (1876), and (*B*) figural tree showing an orangutan, a chimpanzee, a gorilla, and a human, from *Anthropogenie; oder, Entwickelungsgeschichte des Menschen* (1874).

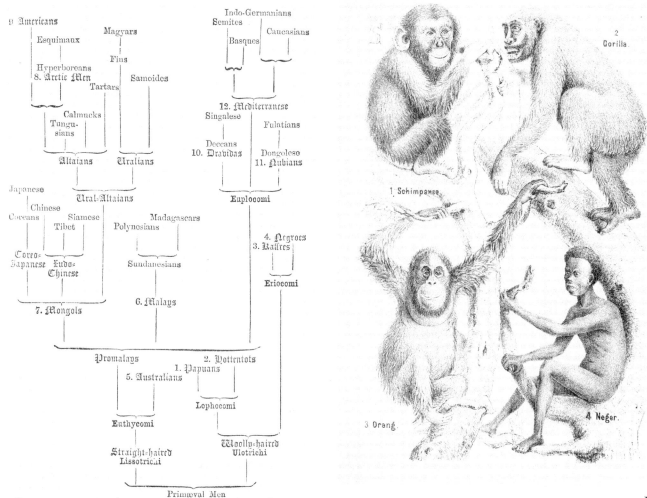

A

B

if quite fancifully and often tinged with racism. This culminates in *Anthropogenie; oder, Entwickelungsgeschichte des Menschen* (*Anthropogeny; or, Evolutionary History of Man*, 1874). It is in this volume that we find Haeckel's famous "Stammbaum des Menschen" (Genealogical Tree of Humans), his "great oak" (see figure 5.9).

In this book as well as his earlier works, Haeckel argues in the text and shows on his trees the origin of humans as having occurred as a singular event, or what we today call monophyly. Yet in the text, he confounds what we know as grades versus clades of evolution, the problem that later scientists such as Alfred Sherwood Romer (1933, 1971) struggled with in deciding between horizontal (grades) or vertical (clades) classification (see figure 6.8*C*). For Haeckel (1874), the best-known and disturbing example comes in a figural tree showing an orangutan, a chimpanzee, a gorilla, and a human labeled "Neger," a German word now often regarded as an ethnic slur (see figure 8.2*B*). In all of Haeckel's trees and elsewhere in his text, he claims monophyly for humans, yet an English translation of the text explaining the diagram reads, "Both the African Manlike Apes [referring in his text to gorillas and chimpanzees] are black in colour, and like their countrymen, the Negroes, have the head long from back to front (dolichocephalic). The Asiatic Man-like Apes are, on the contrary mostly of a brown, or yellowish brown colour, and have the head short from back to front (brachycephalic), like their countrymen, the Malays and Mongols" (1876:180–81).

In its repulsive absurdity, this particular example shows us that Haeckel and some of his contemporaries too often did not realize the inconsistency in confusing grades and clades in evolution. It had a special grip on those biologists such as Haeckel who saw no problems couching racist views under the aegis of science. Even seemingly progressive ideas of improving humankind quickly turned ugly because of value judgments that people necessarily placed on various "desirable" attributes. Such was the case for the concept of eugenics, a term coined by Francis Galton (1822–1911) in *Inquiries into Human Faculty and Its Development* (1883): "We greatly want a brief word to express the science of improving stock, which is by no means confined to questions of judicious mating, but which, especially in the case of man, takes cognisance of all influences that tend in however remote a degree to give to the more suitable races or strains of blood a better chance of prevailing speedily over the less suitable than they otherwise would have had. The word *eugenics* would sufficiently express the idea" (25). For what was to come in the twentieth century, this was chillingly prescient.

Galton was a cousin of Charles Darwin, who for a number of years entertained the misguided idea that he could apply principles of animal breeding to humans. Although at first seemingly well meaning, it fostered some of the most heinous acts against humanity. Accordingly, I find a tree-like symbol used as the logo for the Second International Eugenics Congress, held at the American Museum of Natural History in New York in 1921, especially apropos in its portent (figure 8.3*A*). Henry Fairfield Osborn, president of the American Museum of Natural History, presided at this congress. Haeckel died in 1919, before the founding of the National Socialist Party, and thus we do not know how he might have viewed the ascendance of the Nazis. We do know, however, that Osborn supported both Mussolini's and Hitler's "racial hygiene" programs (Rainger 1991).

The idea of northern European racial superiority permeated not only Osborn's social views but his science as well, in this way not unlike Haeckel. Osborn's book *Men of the Old Stone Age: Their Environment, Life and Art* (1915) fairly represented our knowledge at the time, as well as fairly showing racial attitudes. Osborn's views on the origin of humans, however, found little traction. As discussed in chapter 5, he was no fan of Darwin, instead invoking his neo-Lamarckian aristogenesis, including the inevitability that this unseen force would result in humans. In keeping with these views, Osborn argued that although humans shared ancestry with the apes, the human lineage had a long and separate history stretching back into the Oligocene epoch. In support of this idea, he lauded the importance of the Late Pliocene *Eanthropus dawsoni*, known to posterity as "Piltdown Man." As were many scientists at the time, Osborn was taken in by the Piltdown hoax and even included it on "Ancestral Tree of the Anthropoid Apes and of Man," showing the human lineage dating back to the Oligocene (see figure 8.3*B*). A number of new fossil hominid species as well as new material of known species since Haeckel's work some forty years earlier adorn Osborn's aristogenetic upward march to *Homo sapiens*. These fossil species also occur in the figure that he describes as a "tree showing the main theoretic lines of descent of the chief Pre-Neolithic races discovered in Western Europe" (see figure 8.3*C*). Note the major racial categories "Narrow heads" and "Broad heads" that further subdivide the human lineages. Osborn indicates that these branches of *Homo sapiens* known from the Upper Paleolithic likely separated one from another in the Lower Paleolithic of Asia, likely based on the aforementioned finds that had been made there only a few decade earlier.

Our Split with Chimps: Possibly Older Than We Thought

As discussed in chapter 7, fully twelve years earlier, in 1904, George Nuttall had produced a monographic analysis of blood sera from a variety of animals. His repeated phylogeny by Dubois from 1896 (see figure 7.6*A*) shows a split between apes and human ancestors in the mid-Miocene. Although the timing of the split was not at issue for him, Nuttall indicated that his blood serum work exhibited a similar pattern of splitting among monkeys, apes, and humans. Both the timing of these splits (unlike Osborn's estimates) and the pattern of splitting held fairly constant until about the 1960s. When dramatic ideas of human evolution did arrive, they occurred because of fieldwork, first in southern and then in eastern Africa starting in the 1920s. In 1925, Raymond Dart (1893–1988) described and named *Australopithecus africanus* based on the partial skull of a young individual. It was largely dismissed as an ape by the scientific establishment because it occurred in Africa rather than Europe or Asia, where other hominids were known; its brain size fell

FIGURE 8.3 (*A*) Tree-like symbol used as the logo for the Second International Eugenics Congress, held at the American Museum of Natural History in New York in 1921; Henry Osborn's (*B*) "Ancestral Tree of the Anthropoid Apes and of Man" and (*C*) "tree showing the main theoretic lines of descent of the chief Pre-Neolithic races discovered in Western Europe," from *Men of the Old Stone Age* (1915).

A

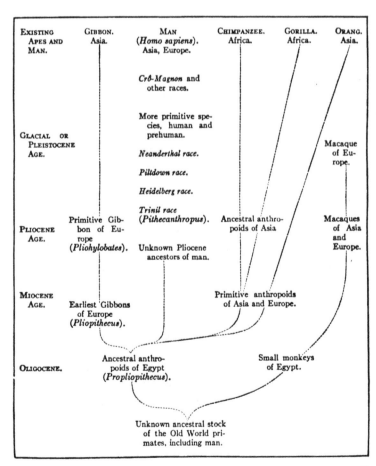

B

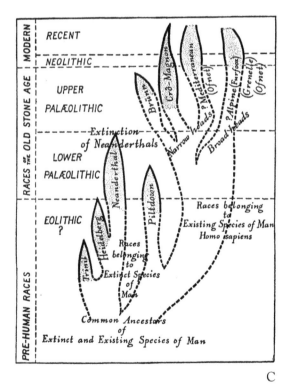

C

in the range of apes, not early humans; and someone outside the scientific inner circle had described it. This all began to change with the recovery of a number of other specimens of australopithecines from South Africa and then in the Rift Valley of East Africa beginning the 1950s, largely through the tenacity of Louis Leakey (1903–1972) and his second wife, Mary (1913–1996). Many paleoanthropologists have followed in the footsteps of Dart in South Africa and the Leakeys in East Africa. Although consensus building among paleoanthropologists may be like herding cats, something like seven to nine species of *Australopithecus* are now generally recognized, ranging in age from about 4 to 2 mya. How they relate to one another; to even earlier possible hominids, dating to perhaps 7 mya; and to the origin of *Homo* remains in dispute. Of interest here is how they have been portrayed in evolutionary trees to answer particularly the questions of when the human and chimpanzee lineages separated from each other and when and how humans migrated from Africa.

When did the chimpanzee and human lineages split? Recall that until 1963, the commonly accepted wisdom held that chimpanzees and gorillas formed a clade with humans as their sister group, such as shown by George Gaylord Simpson (1963) based on fossils, and that this split could be as old as 12 mya (see figure 7.7*B* and *C*). In the same year (and publication), however, Morris Goodman used molecular techniques to argue that these three lineages formed an undifferentiated tritomy (see figure 7.7*A*). Further, in 1967 Vincent Sarich and Allan Wilson, also using molecular techniques, argued that this three-way split dated to about 5 mya (see figure 7.6*B*). In 1984 and again in 1987, Charles Sibley and Jon Ahlquist resolved this tritomy, once again with molecular techniques, to show that chimpanzees and humans were each other's nearest relatives and that the split was between 7.7 and 6.3 mya. In the space of about a hundred years, then, evolutionary trees with humans as the pinnacle of creation changed to trees with humans as the sister clade of chimpanzees. Further, the staggering array of pre-australopithecines, australopithecines, and early *Homo* coming from Africa, even if their relationships could not and still cannot be sorted out, proved Darwin correct: humans' origins were to be found in Africa among their nearest ape relatives.

The newest twist for the split of the chimpanzee and human lineages comes from fossil material as well as molecular data. The often provocative weekly magazine *New Scientist* ran a cover story in November 2012 by one of its editors, Catherine Brahic, with the teasing title "Our True Dawn" and the intriguing argument that humans have been human for a lot longer than we thought. The two accompanying comparative tree figures present the dilemma (figure 8.4*A*).

FIGURE 8.4 (*A*) Old and new views of the split between humans and chimpanzees, with the earlier dates for the fork in the evolutionary road having significant consequences for our understanding of the human family tree, redrawn after Catherine Brahic's "Our True Dawn" (2012); (*B*) Rebecca Cann, Mark Stoneking, and Allan Wilson's U-shaped mitochondrial phylogeny, based on 147 people drawn from five geographic populations, supporting the ancestry from one African woman 200,000 years ago, from "Mitochondrial DNA and Human Evolution" (1987); (*C*) Stoneking's four models for the origin and spread of modern humans, from "Human Origins" (2008). ([*B*] and [*C*] reproduced with permission of the Nature Publishing Group)

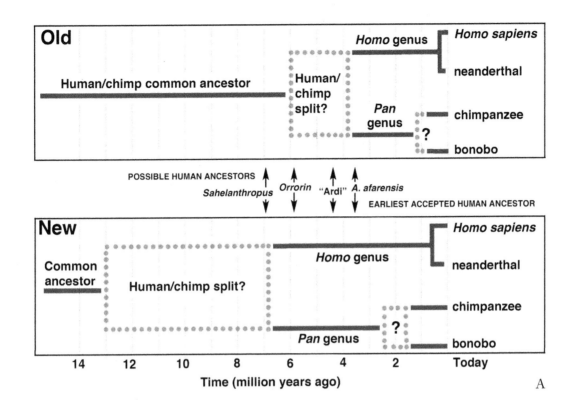

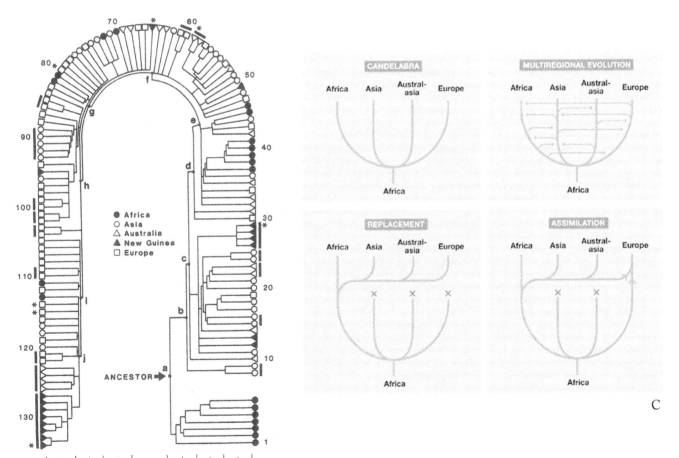

The old view dates the human–chimp split between 6 and 4 mya, based mostly on the molecular data that argued for around seventy-five mutations in genomes between each generation. The species *Australopithecus afarensis* of the fossil Lucy fame, which almost all accept as being a literally upright early representative of the human lineage at almost 4 mya, bangs up against the youngest molecularly based dates for the human–chimp split. But the problem with the molecular data, according to the paleoanthropologists, is that they bracket quite possible to very likely pre-australopithecine members of the human lineage from Central and East Africa back as far as 7 mya (see "possible human ancestors" in figure 8.4*A*). Enter new direct measurements of generational mutation rates of change that may be as low as thirty-six per generation, or half the original seventy-five mutations for older estimates. For both molecular data and fossils, the split now can be pushed back to between 8 and 7 mya, although a minority of paleoanthropologists argue for as early as 13 mya. In the teaser for the article, Brahic writes that the argument for the timing of when humans and chimps split appears to be on the verge of being settled. It certainly makes for good press, but I would not hold my breath, given the perceived importance of the issue.

Leaving Home: A Recent Departure from Africa

The next and more widely known revelation does not entail our deepest roots but our much more recent heritage: When did we depart Africa? And what does this imply about the meaning of human races? A paper published in 1987 by Rebecca Cann, Mark Stoneking, and Allan Wilson proclaimed, "Mitochondrial DNAs from 147 people, drawn from five geographic populations have been analysed by restriction mapping. All these mitochondrial DNAs stem from one woman who is postulated to have lived about 200,000 years ago, probably in Africa. All the populations examined except the African population have multiple origins, implying that each area was colonised repeatedly" (31). So there was an Eve after all, and she hailed from Africa 200,000 years ago. Although the paper did not say it specifically, the implication emerged that all living humans could be traced to this one woman, or at least to the population to which she belonged. Being based on mitochondrial DNA, which is inherited only through the maternal lineage, the study could not say from where the father hailed. This first attempt is shown in figure 8.4*B* in a nice upside-down U-shaped but still discernible branching tree. One of the more curious aspects of this tree is how the different geographic groups cluster, or rather do not cluster in many instances. None of the geographic groups forms a cluster totally separate from any other. Cann and her colleagues found that the tree first breaks into one branch that includes only Africans, whereas the other branch includes Africans as well as non-Africans (see figure 8.4*B*), which leads the authors to conclude that the common mitochondrial ancestor was African. What does this say about the usually perceived racial groupings based on aspects of skin color, hair texture, skull shape, and facial features? This study and others the authors cite indicate that various molecularly based differences can be greater within usually identified races than between them. Some African maternal lineages share a more recent ancestry with non-Africans than with Africans, as the authors' tree suggests. Thus much of what we attribute to race might be inferred more as holdovers of some wider ancestral traits and less from those with whom we share more recent ancestry.

The results of this and similar studies did not escape critics, most notably those who supported what came to be called the Multiregional Hypothesis (for example, Wolpoff et al. 1988) for the origin of modern humans versus what became known as the Out of Africa or African Eve Hypothesis. Before the latter hypothesis appeared, the unchallenged idea was that *Homo erectus* originated in Africa some 2 mya, subsequently spreading throughout much of the Old World. Following this, these archaic forms remained an interbreeding population, and through gene flow and local adaptation this Old World population evolved from *Homo erectus* to archaic *Homo sapiens* and then to modern humans. In the late 1960s and into the early 1970s, some even thought that the australopithecines constituted only one or a few species (for example, Wolpoff 1968) that fit into a neat straight-line evolutionary portrayal, such as Rudolph Zallinger presented in F. Clark Howell's *Early Man* (1965) (see figure 1.8*A*). This soon disabused idea melted away as more and more different australopithecines turned up in the fossil record. Even as evidence mounted, we could not rid ourselves of the idea that we represent this singular evolutionary progression rather than the reality that we are but one twig on the tree, even though most of the rest had been hacked away by extinction—possibly through some of our agency.

Out of the numerous hypotheses for the origin and spread of modern humans, Stoneking (2008) presents succinct tree models for the four most clearly articulated hypotheses. He identifies these as the candelabra, multiregional evolution, replacement, and assimilation models (see figure 8.4*C*). Stoneking regards the candelabra as perhaps the oldest model, which prevailed for decades. According to this model, the common human ancestor dates back to some 2 mya, with modern humans arising separately and probably in considerable isolation in Africa, Europe, Asia, and Australasia. As Stoneking notes, this view helped foster racist ideas. He identifies specifically Carleton Coon (1904–1981), but the even earlier ideas of Osborn fall within this hypothetical framework. Within this racist scenario, Europeans had a longer time in which to evolve, whereas Africans were the last to do so.

Stoneking's three other models share the now strong evidence that until between 2 and 1.5 mya, human evolution took place in Africa. Following this, ancestral humans began a series of migrations throughout the Old World, but when and how they occurred is in dispute. In the multiregional evolution model, the ancestral humans left Africa and spread out in the Old World, but as Stoneking shows by the small horizontal arrows, considerable migration and gene flow occurred during this long interval so that throughout the Old World humans evolved in concert, although local conditions certainly caused differing adaptational modifications (see figure 8.4*C*, *upper right*). Stoneking's replacement model, by contrast, argues that modern humans arose some 300,000 to 200,000 years ago, spreading out of Africa between 100,000 and 50,000 years ago and replacing the local populations as they migrated (see figure 8.4*C*, *lower left*). This represents one form of the Out of Africa model. The fourth and final model, assimilation, differs in that it argues that some evidence suggests that the process did not involve solely replacement, but that as modern humans spread from Africa, they on occasion interbred with indigenous archaic humans (see figure 8.4C, *lower right*). A study showing that non-Africans share about 2 to 3 percent of their genome with Neanderthals (Currat and Excoffier 2011) bolsters the idea that as modern humans migrated out of Africa, they may have mated with Neanderthals. Other explanations suggest that this small percentage may be a genomic holdover that was shared by the last Neanderthal and human ancestor before that common ancestor departed from Africa, even before the more recent exodus of modern humans. Whichever, if any, of these hypotheses proves correct, the evidence that we can hang on the family tree shows that, in a very real sense, we are all Africans.

As these four hypotheses show, evolution has not produced a nice neat phylogeny of humans or, for that matter, any group of plants or animals resulting in a neatly branching family tree. Evolution, as with life in general, is complex and inconveniently messy. When viewed backward through a telescope of geologic time, evolutionary patterns might seem discrete episodes of splitting and resplitting with the twain never meeting, but even our best, albeit myopic, visual metaphors are just that—metaphors.

References

Agassiz, L. J. R. 1833–1844. *Recherches sur les poissons fossiles*. Vol. 1. Neuchâtel: Imprimerie de Petitpierre et Prince.

Agassiz, L., and A. A. Gould. 1848. *Principles of Zoölogy: Touching the Structure, Development, Distribution, and Natural Arrangement of the Races of Animals, Living and Extinct*. Boston: Gould, Kendall and Lincoln.

Ahlquist, J. 1999. Charles G. Sibley: A commentary on 30 years of collaboration. *The Auk* 116:856–60.

Alejos-Grau, C. J. 1994. *Diego Valadés, educador de la Nueva España: Ideas pedagógicas de la Rhetorica christiana (1579) (Acta philosophica)*. Pamplona: Ediciones Eunate.

Anderson, L. 1976. Charles Bonnet's taxonomy and chain of being. *Journal of the History of Ideas* 37:45–58.

Appel, T. A. 1980. Henri de Blainville and the animal series: A nineteenth-century Chain of Being. *Journal of the History of Biology* 13:291–319.

Archibald, J. D. 2009. Edward Hitchcock's pre-Darwin (1840) "Tree of life." *Journal of the History of Biology* 42:561–92.

———. 2011. *Extinction and Radiation: How the Fall of the Dinosaurs Led to the Rise of the Mammals*. Baltimore: Johns Hopkins University Press.

———. 2012. Darwin's two competing phylogenetic trees: Marsupials as ancestors or sister taxa? *Archives of Natural History* 39:217–33.

Archibald, J. D., and D. Deutschman. 2001. Quantitative analysis of the timing of origin of extant placental orders. *Journal of Mammal Evolution* 8:107–24.

Archibald, J. D., Y. Zhang, T. Harper, and R. L. Cifelli. 2011. *Protungulatum*, confirmed Cretaceous occurrence of an otherwise Paleocene eutherian (placental?) mammal. *Journal of Mammalian Evolution* 18:153–61.

Argyll, G. D. Campbell, Duke of. 1867. *The Reign of Law*. London: Alexander Strahan.

Aristotle. 1991. *History of Animals*. Vol. 3, books 7–10. Edited and translated by D. M. Balme. Loeb Classical Library 439. Cambridge, Mass.: Harvard University Press.

———. 2007. *The History of Animals.* Translated by D. W. Thompson. Adelaide: eBooks@ Adelaide. http://ebooks.adelaide.edu.au/a/aristotle/history/index.html.

Arnar, A. S. 1990. *Encyclopedism from Pliny to Borges.* Chicago: University of Chicago Library.

Augier, A. 1801. *Essai d'une nouvelle Classification des Végétaux conforme à l'Ordre que la Nature paroit avoir suivi dans le Règne Végétal: d'ou Resulte une Méthode qui conduit à la Conaissance des Plantes & de leur Rapports naturels.* Lyon: Bruyset Aîné.

Barbançois, C.-H. de. 1816. Observations sur la filiation des animaux, depuis le polype jusqu au singe. *Journal de Physique de Chimie et d'Histoire naturelle* 82:444–48.

Barsanti, G. 1992. *La Scala, la Mappa, l'Albero: Immagini e classificazioni della natura fra Sei e Ottocento.* Florence: Sansoni Editore.

Begun, D. R. 2009. *Dryopithecus,* Darwin, de Bonis, and the European origin of the African apes and the human clade. *Geodiversitas* 31:789–816.

Bentham, G. 1873. Notes on the classification, history, and geographical distribution of Compositae. *Journal of the Linnean Society of London* 13:335–577.

Bettini, M. 1991. *Anthropology and Roman Culture: Kinship, Time, Images of the Soul.* Translated by J. Van Sickle. Baltimore: Johns Hopkins University Press.

Biblia Sacra Vulgata (Vulgate): Holy Bible in Latin (Latin ed.). Ca. 382 C.E.

Bigoni, F., and G. Barsanti. 2011. Evolutionary trees and the rise of modern primatology: The forgotten contribution of St. George Mivart. *Journal of Anthropological Sciences* 89:1–15.

Bininda-Emonds, O. R. P., M. Cardillo, K. E. Jones, R. D. E. MacPhee, R. M. D. Beck, R. Grenyer, S. A. Price, R. A. Vos, J. L. Gittleman, and A. Purvis. 2007. The delayed rise of present-day mammals. *Nature* 446:507–12.

Blumenbach, J. F. 1779. *Handbuch der Naturgeschichte.* Göttingen: Dieterich.

Bonnet, C. 1745. *Traité d'Insectologie; ou observations sur les pucerons:* Pt. 1; *ou observations sur quelques especes de vers d'eau douce, qui coupés par morceaux, deviennent autant d'animaux complets:* Pt. 2. Paris: Durand.

———. 1764. *Contemplation de la nature.* Vol. 1. Amsterdam: Marc-Michel Rey.

———. 1781. *Oeuvres d'histoire naturelle et de philosophie.* Vol. 4, pt. 1, *Contemplation de la nature.* Quarto ed. Neuchâtel: S. Fauche.

Bowler, P. J. 1988. *The Non-Darwinian Revolution.* Baltimore: Johns Hopkins University Press.

———. 1989. *Evolution: The History of an Idea.* Rev. ed. Berkeley: University of California Press.

Boyle, L. 1989. *A Short Guide to St. Clement's Rome.* Rome: Collegio San Clemente.

Brahic, C. 2012. Our true dawn: Pinning down human origins. *New Scientist* 216:34–37.

Bronn, H. G. 1837–1838. *Lethaea geognostica, oder Abbildungen und Beschreibungen der für die Gebirgs-Formationen bezeichnendsten Versteinerungen.* Stuttgart: E. Schweizerbart's Verlagshandlung.

———. 1850–1856. *Lethaea geognostica, oder Abbildungen und Beschreibungen der für die Gebirgs-Formationen bezeichnendsten Versteinerungen.* 3rd ed. Stuttgart: E. Schweizerbart's Verlagshandlung.

———. 1858. *Untersuchungen über die Entwicklungs-Gesetze der organischen Welt während der Bildungszeit unserer Erd-Oberfläche.* Stuttgart: E. Schweizerbart'sche Verlagshandlung und Druckerei.

———. 1860. Schlusswort des Übersetzers. In *Über die Entstehung der Arten im Thier- und Pflanzen-Reich durch natürliche Züchtung; oder, Erhaltung der vervollkommneten Rassen im Kampfe um's Daseyn,* by Charles Darwin. Stuttgart: Schweizerbart'sche Verlagshandlung.

———. 1861. Essai d'une résponse à la question de prix proposée en 1850. *Supplément aux Comptes Rendus des Séances de l'Academie des Sciences Paris* 2:377–918.

Brown, W. H. 1890. "Dates of Publication of 'Recherches sur les Poissons Fossiles' . . . par L. Agassiz." In *A Catalogue of British Fossil Vertebrata*, edited by A. S. Woodward and C. D. Sherborn, xxv–xxix. London: Dulau.

Browne, J. 1983. *The Secular Ark: Studies in the History of Biogeography*. New Haven, Conn.: Yale University Press.

———. 2002. *Charles Darwin: The Power of Place*. New York: Knopf.

Brundin, L. 1965. On the real nature of Transantarctic relationships. *Evolution* 19:496–505.

———. 1966. Transantarctic relationships and their significance as evidenced by chirnonmid midges. *Kungliga Svenksa Vetenskapsakademiens Handlingar*, 4th ser., 11, no. 1:1–472.

Brush, A. H. 2003. Charles Gald Sibley, 1917–1990. *National Academy of Sciences Biographical Memoirs* 83:1–24.

Burkhardt, R. W., Jr. 1980. The *Zoological Philosophy* of J. B. Lamarck. In *Zoological Philosophy: An Exposition with Regard to the Natural History of Animals*, by J. B. Lamarck, translated by H. Elliot, xv–xxxix. Chicago: University of Chicago Press.

Burns, K. J., S. J. Hackett, and N. K. Klein. 2002. Phylogenetic relationships and morphological diversity in Darwin's finches and their relatives. *Evolution* 56:1240–52.

Butts, S. 2010. A general view of the Animal Kingdom conservation of the Anna Maria Redfield wall chart. *Yale Environmental News* 16:10–11.

Cambridge University Library. 1961. *Darwin Library. List of Books Received in the University Library Cambridge, March–May 1961*. Pamphlet. Cambridge: Cambridge University Library.

Caneva, G. 2010. *The Augustus Botanical Code: Ara Pacis: Speaking to the People Through the Images of Nature*. Rome: Gangemi Editore.

Cann, R. L., M. Stoneking, and A. C. Wilson. 1987. Mitochondrial DNA and human evolution. *Nature* 325:31–36.

Carpenter, W. B. 1839. *Principles of General and Comparative Physiology, Intended as an Introduction to the Study of Human Physiology, and as a Guide to the Philosophical Pursuit of Natural History*. London: John Churchill.

———. 1841. *Principles of General and Comparative Physiology, Intended as an Introduction to the Study of Human Physiology, and as a Guide to the Philosophical Pursuit of Natural History*. 2nd ed. London: John Churchill.

———. 1854. *Principles of Comparative Physiology*. A new American edition, from the 4th and rev. London ed. Philadelphia: Blanchard and Lea.

Carus, P. 1900. *The History of the Devil and the Idea of Evil, from the Earliest Times to the Present Day*. Chicago: Open Court.

Cavalli-Sforza, L. L., and A. W. F. Edwards. 1965. Analysis of human evolution. In *Genetics Today, Proceedings of the XI International Congress of Genetics, The Hague, the Netherlands, September 1963*, edited by S. J. Geerts, 3:923–33. Oxford: Pergamon Press.

Chamberlain, J. L. 1900. *Universities and Their Sons*. Boston: Herndon.

[Chambers, R.] 1844. *Vestiges of the Natural History of Creation*. London: John Churchill.

Chesnova, L. V. 1972. Razvitiye filogyenyetichyeskoi sistyematiki zhivotnih [Development of phylogentic animal taxonomy]. In *Istoriya Biologii s Dryevnyeyshih Vryemyen do Nachala XX Veka* [History of biology from ancient times to the beginning of the twentieth century], edited by S. R. Mikulinskii, 344–53. Moscow: Nauka.

Churches of Rome. 2012. San Clemente. http://romanchurches.wikia.com/wiki/San_Clemente.

Coggon, J. 2002. Quinarianism After Darwin's "Origin": The Circular System of William Hincks. *Journal of the History of Biology* 35:5–42.

Colbert, E. H. 1975. William King Gregory, 1876–1970. *National Academy of Sciences Biographical Memoirs* 46:91–133.

———. 1982. Alfred Sherwood Romer, 1894–1973. *National Academy of Sciences Biographical Memoirs* 53:265–94.

Collegio S. Clemente, ed., and M. Gerardi. 1992. *San Clemente Roma*. Marano Vicentino: KinaItalia/EuroGrafica.

Cook, R. 1988. *The Tree of Life: Image for the Cosmos*. New York: Thames and Hudson.

Cooper, A., and D. Penny. 1997. Mass survival of birds across the Cretaceous–Tertiary boundary: Molecular evidence. *Science* 275:1109–13.

Cope, E. D. 1868. On the origin of genera. *Proceedings of the Academy of Natural Sciences of Philadelphia* 20:242–300.

———. 1871. The method of creation of organic forms. *Proceedings of the American Philosophical Society* 12:229–63.

———. 1884. *The Vertebrata of the Tertiary Formation of the West*. Book 1. Report of the United State Geological Survey of the Territories, vol. 3. F. V. Hayden, Geologist-in-charge. Washington, D.C.: Government Printing Office.

———. 1887. *The Origin of the Fittest: Essay of Evolution*. New York: Appleton.

———. 1896. *The Primary Factors of Organic Evolution*. Chicago: Open Court.

Crow, C. 2006. The Ara Pacis. *History Today* 56:5.

Currat, M., and L. Excoffier. 2011. Strong reproductive isolation between humans and Neanderthals inferred from observed patterns of introgression. *Proceedings of the National Academy of Sciences* 108:15129–34.

Cuvier, G. 1812. Sur un nouveau rapprochement à établir entre les classes qui composent le règne animal. *Annales du Muséum National Histoire Naturelle* 19:73–84.

———. 1817. *Le Règne animal distribué d'après son organisation, pour servir de base à l'histoire naturelle des animaux et d'introduction à l'anatomie compare*. 4 vols. Paris: Deterville.

———. 1825. Nature. In *Dictionnaire des Sciences Naturelles*, edited by F. G. Levrault, 34:261–68. Strasbourg: Le Normant.

Dart, R. 1925. *Australopithecus africanus*: The man-ape of South Africa. *Nature* 115:195–99.

Darwin, C. R., ed. 1838. *Birds Part 3 No. 1 of The Zoology of the Voyage of H.M.S. Beagle by John Gould. Edited and Superintended by Charles Darwin*. London: Smith, Elder.

———, ed. 1838–1843. *The Zoology of the Voyage of H.M.S. Beagle, Under the Command of Captain Fitzroy, R.N., During the Years 1832 to 1836*. 5 vols. London: Smith, Elder.

———. 1839. *Journal and Remarks, 1832–1836*. Vol. 3 of *Narrative of the Surveying Voyages of His Majesty's Ships Adventure and Beagle, Between the Years 1826 and 1836, Describing Their Examination of the Southern Shores of South America and the Beagle's Circumnavigation of the Globe*. Edited by R. FitzRoy. London: Henry Colburn.

———. 1859. *On the Origin of Species by Means of Natural Selection, or the Preservation of Favoured Races in the Struggle for Life*. London: John Murray.

———. 1860. *Über die Entstehung der Arten im Thier- und Pflanzen-Reich durch natürliche Züchtung; oder, Erhaltung der vervollkommneten Rassen im Kampfe um's Dasey*. Translated by H. G. Bronn. Based on the 2nd ed. Stuttgart: Schweizerbart'sche Verlagshandlung.

———. 1868. *The Variation of Animals and Plants Under Domestication*. 2 vols. London: John Murray.

———. 1871. *The Descent of Man, and Selection in Relation to Sex*. 2 vols. London: John Murray.

———. 1960. Darwin's notebooks on the transmutation of species. Part 1, First Notebook (July 1837–February 1838). Edited by G. de Beer. *Bulletin of the British Museum (Natural History) Historical Series* 2:27–73.

———. 1975. *Charles Darwin's Natural Selection: Being the Second Part of His Big Species Book Written from 1856 to 1858*. Edited by R. C. Stauffer. Cambridge: Cambridge University Press.

———. 1987. *The Works of Charles Darwin*. Vol. 5, *The Zoology of the Voyage of H.M.S. Beagle*, Pt. 3, *Birds*. Edited by P. H. Barrett and R. B. Freeman. New York: New York University Press.

———. 2005. *The Correspondence of Charles Darwin*. Vol. 15, *1867*. Edited by F. Burkhardt, D. M. Porter, S. A. Dean, S. Evans, S. Innes, A. M. Pearn, A. Sclater, and P. White. Cambridge: Cambridge University Press.

———. 2008. *The Correspondence of Charles Darwin*. Vol. 16, pt. 1, *January–June 1868*. Edited by F. Burkhardt, J. A. Secord, S. A. Dean, S. Evans, S. Innes, A. M. Pearn, and P. White. Cambridge: Cambridge University Press.

———. 2010. *The Correspondence of Charles Darwin*. Vol. 18, *1870*. Edited by F. Burkhardt, J. A. Secord, S. A. Dean, S. Evans, S. Innes, A. M. Pearn, and P. White. Cambridge: Cambridge University Press.

———. Charles Darwin Papers. Darwin Manuscripts Project. http://darwin.amnh.org/browse.php?mode=uc&pid=72001

Darwin, E. 1803. *The Temple of Nature; or, the Origin of Society*. London: J. Johnson.

Dayrat, B. 2003. The roots of phylogeny: How did Haeckel build his trees? *Systematic Biology* 52:515–17.

Desmond, A. 1989. *The Politics of Evolution: Morphology, Medicine, and Reform in Radical London*. Chicago: University of Chicago Press.

———. 1994. *Huxley: The Devil's Disciple*. London: Michael Joseph.

———. 1997. *Huxley: Evolution's High Priest*. London: Michael Joseph.

Dio Cassius. 1917. *Roman History*. Vol. 6, books 51–55. Translated by E. Cary and H. B. Foster. Loeb Classical Library 83. Cambridge, Mass.: Harvard University Press.

Doolittle, W. F. 1999. Phylogenetic classification and the universal tree. *Science* 284:2124–28.

dos Reis, M., P. C. J. Donoghue, and Z. Yang. 2014. Neither phylogenomic nor palaeontological data support a Palaeogene origin of placental mammals. *Biology Letters* 10, no. 1:1–4.

Dubois, E. 1896. *Pithecanthropus erectus*, eine Stammform des Menschen. *Anatomischer Anzeiger* 12:1–22.

Eichwald, E. 1829. *Zoologia specialis quam expositis animalibus tum vivis, tum fossilibus potissimum Rossiae in universum, et Poloniae in species, in usum lectionum publicarum in Universitate Caesarea Vilnensi habendarum. Pars prior. Propaedeuticam zoologiae atque specialem Heterozoorum expositionem continens*. Vilnae: Josephus Zawadzki.

Eigen, E. D. 1997. Overcoming first impressions: Georges Cuvier's types. *Journal of the History of Biology* 30:179–209.

Eldredge, N., and J. Cracraft. 1980. *Phylogenetic Patterns and the Evolutionary Process: Method and Theory in Comparative Biology*. New York: Columbia University Press.

Eldredge, N., and S. J. Gould. 1972. Punctuated equilibria: An alternative to phyletic gradualism. In *Models in Paleobiology*, edited by T. J. M. Schopf, 82–115. San Francisco: Freeman, Cooper.

Fane, D. 1997. Exhibition review. *Colonial Latin American Review* 6:243–51.

Faust, L. 1585 (1586). *Anatomia statuae Danielis*. Leipzig.

Feduccia, A. 1995. Explosive evolution in Tertiary birds and mammals. *Science* 267:637–38.

———. 1996. *The Origin and Evolution of Birds*. New Haven, Conn.: Yale University Press.

Fleischer, G. 1802. *Annuaire de la librairie: Première année*. Paris: Levrault.

Foucault, M. 1970. *The Order of Things: An Archaeology of the Human Sciences*. New York: Pantheon Books.

Freeman, R. B. 1977. *The Works of Charles Darwin: An Annotated Bibliographical Handlist*. 2nd ed. London: Dawson Archon Books.

Friedman, W. E. 2009. The meaning of Darwin's "abominable mystery." *American Journal of Botany* 96:5–21.

Fürbinger, M. 1870. *Die Knochen und Muskeln der Extremitäten bei den schlangenähnlichen Sauriern*. Leipzig: Verlag von Wilhelm Engelmann.

———. 1888. *Untersuchungen zur Morphologie und Systematik der Vögel, zugleich ein Beitrag zur Anatomie der Stütz- und Bewegungsorgane*. 2 vols. Amsterdam: T. van Holkema.

Futuyma, D. J. 1986. *Evolutionary Biology*. 2nd ed. Sunderland, Mass.: Sinauer.

Gagarin, M., ed. 2009. *The Oxford Encyclopedia of Ancient Greece and Rome*. Vol. 1. Oxford: Oxford University Press.

Galton, F. 1883. *Inquiries into Human Faculty and Its Development*. London: Macmillan.

Gardiner, B. G. 1982. Tetrapod classification. *Zoological Journal of the Linnean Society* 74:207–32.

Gaudry, J. A. 1862–1867. *Animaux fossiles et géologie de l'Attique, d'après les recherches faites en 1855, 1856 et en 1860*. Paris: F. Savy.

———. 1866. *Considérations générales sur les animaux fossiles de Pikermi*. Paris: F. Savy.

Gerardi, M. 1988. *Mosaico di S. Clemente*. Milan: Kina Italia.

Germano di San Stanislao, P. 1894. *La Casa Celimontana dei SS. Martiri Giovanni e Paolo*. Rome: Tipografica della Pace di F. Cuggiani.

Gianfreda, G. 2008. *Il mosaico di Otranto: Biblioteca medioevale in immagini*. Lecce: Edizioni del Grifo.

Glangeaud, P. 1910. Albert Gaudry and the evolution of the animal kingdom. *Smithsonian Report for 1909* 1969:417–29.

Gliboff, S. 2007. H. G. Bronn and the history of nature. *Journal of the History of Biology* 40:259–94.

———. 2008. *H. G. Bronn, Ernst Haeckel, and the Origins of German Darwinism: A Study in Translation and Transformation*. Cambridge, Mass.: MIT Press.

Goodman, M. 1963. Man's place in the phylogeny of the primates as reflected in serum proteins. In *Classification and Human Evolution*, edited by S. L. Washburn, 204–34. Chicago: Aldine.

Goodman, M., G. W. Moore, and G. Matsuda. 1975. Darwinian evolution in the genealogy of haemoglobin. *Nature* 253:603–8.

Gould, S. J. 1977. *Ontogeny and Phylogeny*. Cambridge, Mass.: Belknap Press of Harvard University Press.

———. 1989. *Wonderful Life: The Burgess Shale and the Nature of History*. New York: Norton.

———. 1991. *Bully for Brontosaurus: Reflections in Natural History*. New York: Norton.

———. 1993. *Eight Little Piggies: Reflections in Natural History*. New York: Norton.

Green, R., M. Evans, C. Bischoff, and M. Curschmann. 1979. *Hortus Deliciarum of Herrad of Hohenbourg, a Reconstruction*. 2 vols. London: Warburg Institute.

Greenblatt, S. 2011. *The Swerve: How the World Became Modern*. New York: Norton.

Gregory, W. K. 1951. *Evolution Emerging*. 2 vols. New York: Macmillan.

Grindon, L. H. 1863. *Life: Its Nature, Varieties, and Phenomena*. 3rd ed. London: F. Pitman.

Gross, B. L., and L. H. Rieseberg. 2005. The ecological genetics of homoploid hybrid speciation. *Journal of Heredity* 96:241–52.

Grünbaum, A. S. F. 1902. Notes on the "blood relationship" of man and the anthropoid apes. *Lancet*, January 18, 143.

Hackett, S. J., R. T. Kimball, S. Reddy, R. C. K. Bowie, E. L. Braun, M. J. Braun, J. L. Chojnowski, W. A. Cox, K.-L. Han, J. Harshman, C. J. Huddleston, B. D. Marks, K. J. Miglia, W. S. Moore, F. H. Sheldon, D. W. Steadman, C. C. Witt, and T. Yuri. 2008. A phylogenomic study of birds reveals their evolutionary history. *Science* 320:1763–68.

Haeckel, E. 1862. *Die Radiolarien (Rhizopoda Radiaria): Eine Monographie*. 2 vols. Berlin: Georg Reimer.

——. 1866. *Generelle Morphologie der Organismen: allgemeine Grundzüge der organischen Formen-Wissenschaft, mechanisch begründet durch die von Charles Darwin reformirte Descendenz-Theorie*. 2 vols. Berlin: Georg Reimer.

——. 1868. *Natürliche Schöpfungsgeschichte: Gemeinverständliche wissenschaftliche Vorträge über die Entwickelungslehre im allgemeinen und diejenige von Darwin, Goethe und Lamarck im Besonderen, über die Anwendung derselben auf den Ursprung des Menschen und andere damit zusammenhängende Grundfragen der Naturwissenschaft*. Berlin: Georg Reimer.

——. 1870. *Natürliche Schöpfungsgeschichte: Gemeinverständliche wissenschaftliche Vorträge über die Entwickelungslehre im allgemeinen und diejenige von Darwin, Goethe und Lamarck im Besonderen, über die Anwendung derselben auf den Ursprung des Menschen und andere damit zusammenhängende Grundfragen der Naturwissenschaft*. 2nd ed. Berlin: Georg Reimer

——. 1873. *Natürliche Schöpfungsgeschichte: Gemeinverständliche wissenschaftliche Vorträge über die Entwickelungslehre im allgemeinen und diejenige von Darwin, Goethe und Lamarck im Besonderen*. 5th ed. Berlin: Georg Reimer.

——. 1874. *Anthropogenie; oder, Entwickelungsgeschichte des Menschen*. Leipzig: Engelmann.

——. 1876. *The History of Creation: Or the Development of the Earth and Its Inhabitants by the Action of Natural Causes*. 2 vols. Translated and revised by E. R. Lankester. New York: Appleton.

Hagen, J. 2004. Interview with Morris Goodman. History of Recent Science & Technology. The Dibner Institute for the History of Science and Technology. http://authors.library.caltech.edu/5456/1/hrst.mit.edu/hrs/evolution/public/goodman.html.

Harshman, J., E. L. Braun, M. J. Braun, C. J. Huddleston, R. C. K. Bowie, J. L. Chojnowski, S. J. Hackett, K.-L. Han, R. T. Kimball, B. D. Marks, K. J. Miglia, W. S. Moore, S. Reddy, F. H. Sheldon, D. W. Steadman, S. J. Steppan, C. C. Witt, and T. Yuri. 2008. Phylogenomic evidence for multiple losses of flight in ratite birds. *Proceedings of the National Academy of Sciences* 105:13462–67.

Heilbron, J. L. 1990. The measure of enlightenment. In *The Quantifying Spirit in the Eighteenth Century*, edited by T. Frängsmyr, J. L. Heilbron, and R. E. Rider, 207–43. Berkeley: University of California Press.

Hennig, W. 1950. *Grundzüge einer Theorie der phylogenetischen Systematik*. Berlin: Deutscher Zentralverlag.

——. 1966 (1979). *Phylogenetic Systematics*. Edited and translated by D. Davis and R. Zangerl. Urbana: University of Illinois Press.

Hestmark, G. 2000. Temptations of the tree. *Nature* 408:911.

Hilgendorf, F. 1863. Beiträge zur Kenntnis des Süßwasserkalkes von Steinheim. Ph.D. diss., University of Tübingen.

——. 1866. Planorbis multiformis *im Steinheimer Süßwasserkalk: Ein Beispiel von Gestaltveränderung im Laufe der Zeit*. Berlin: W. Weber.

——. 1867. Über *Planorbis multiformis* im Steinheimer Süßwasserkalk. *Monatsberichte der Königlich Preussischen Akademie der Wissenschaften zu Berlin* 1866:474–504.

Hillis, D. M., D. Zwickl, and R. Gutell. *Tree of Life*. http://www.zo.utexas.edu/faculty/antisense/DownloadfilesToL.html.

Hitchcock, E. 1840. *Elementary Geology*. Amherst, Mass.: J. S. & C. Adams.

——. 1841. *Elementary Geology*. 2nd ed. Amherst, Mass.: J. S. & C. Adams.

——. 1842. *Elementary Geology*. 3rd ed. New York: Dayton & Newman.

——. 1847. *Elementary Geology*. 8th ed. New York: Mark H. Newman.

——. 1852. *Elementary Geology*. 8th ed. New York: Newman & Ivison.

——. 1855. *Elementary Geology*. 25th ed. New York: Ivision & Phinney.

——. 1856. *Elementary Geology*. 30th ed. New York: Ivison & Phinney.

Hitchcock, E., and C. Hitchcock. 1860. *Elementary Geology*. 31st new ed. New York: Ivison, Phinney.

——. 1862. *Elementary Geology*. New ed. New York: Ivison, Phinney.

——. 1866. *Elementary Geology*. New ed. New York: Ivison, Phinney.

Holmes, O. W. 2006. The enlightenment and early romantic concepts of nature and the self. *Analecta Husserliana* 90:147–75.

Holmes, S. J. 1947. K. E. von Baer's perplexities over evolution. *Isis* 37:7–14.

The Holy Bible (King James Bible). 1611. London: Robert Barker.

Hopwood, N., S. Schaffer, and J. Secord. 2010. Seriality and scientific objects in the nineteenth century. *History of Science* 48:251–85.

Horner, W. B., and M. Leff. 1995. *Rhetoric and Pedagogy: Its History, Philosophy, and Practice: Essays in Honor of James J. Murphy.* Mahwah, N.J.: Erlbaum.

Howell, F. C. 1965. *Early Man.* New York: Time-Life.

Hoyer, B. H., N. W. van de Velde, M. Goodman, and R. B. Roberts. 1972. Examination of hominid evolution by DNA sequence homology. *Journal of Human Evolution* 1:645–49.

Hughes, R. 2011. *Rome: A Cultural, Visual, and Personal History.* New York: Knopf.

Humboldt, A. von 1817. *De distributione geographica plantarum.* Paris: Libraria Graeco-Latino-Germanica.

Hunter, G. W. 1914. *A Civic Biology: Presented in Problems.* New York: American Book.

——. 1926. *A New Civic Biology: Presented in Problems.* New York: American Book.

Huxley, T. H. 1863. *Evidence as to Man's Place in Nature.* London: Williams & Norgate.

——. 1877. *Lectures and Essays.* London: Macmillan.

Isidore, S., Arzobispo de Sevilla. 1483. *Etymologiae: De summo bono.* Venice: Peter Löslein.

——. 2011. *The Etymologies of Isidore of Seville.* Edited and translated by S. A. Barney, W. J. Lewis, J. A. Beach, and O. Berghof. Cambridge: Cambridge University Press.

Klapisch-Zuber, C. 1991. The genesis of the family tree. *I Tatti Studies: Essays in the Renaissance* 4:105–29.

——. 2000. *L'ombre des ancêtres: Essai sur l'imaginaire médiéval de la parenté.* Paris: Fayard.

Kovalevskii (Kowalewky), V. O. 1876. Monographie der Gattung Anthracotherium Cuv. (Three parts, 1873–1874.) *Palaeontographica: Beiträge zur Naturgeschichte der Vorzeit* 22:131–398.

Kren, T. 2009. *Illuminated Manuscripts of Germany and Central Europe in the J. Paul Getty Museum.* Los Angeles: J. Paul Getty Museum.

Kuznetsov, N. I. 1896. Subgenus *Eugentiana* Kusnez. generis *Gentiana* Tournef. *Acta Horti Petropolitana* 15:1–507.

Lacordaire, J. T. 1839–1840. On the geographical distribution of insects. *Edinburgh New Philosophical Journal* 28:170–83.

Lamarck J.-B. 1778 (1779). *Flore françoise ou description succincte des toutes les plantes qui croissant naturellement en France, disposée selon une nouvelle méthode d'analyse à laquelle on a joint la citation de leurs vertus les moins équivoques en médecine, et de leur utilité dans les arts.* Vols. 1 and 2. Paris: Imprimerie Royale.

———. 1786. *Encyclopédie méthodique: Botanique*. Vol. 2. Paris: Panckoucke.

———. 1802. *Recherches sur l'organisation des corps vivans et particulièrement sur son origine, sur la cause de ses développemens et des progrès de sa composition, et sur celle qui, tendant continuellement à la détruire dans chaque individu, amène nécessairement sa mort; précédé du discours d'ouverture du cours de zoologie, donné dans le Muséum national d'Histoire Naturelle*. Paris: Maillard.

———. 1809. *Philosophie zoologique, ou exposition des Considérations relatives à l'histoire naturelle des Animaux; à la diversité de leur organisation et des facultés qu'ils en obtiennent*. Paris: Dentu.

Laporte, L. F. 2000. *George Gaylord Simpson: Paleontologist and Evolutionist*. New York: Columbia University Press.

Larson, E. J. 1997. *Summer for the Gods: The Scopes Trial and America's Continuing Debate over Science and Religion*. New York: Basic Books.

Lartet, É. 1856. Note sur un grand Singe fossile qui se rattache au groupe des Singes supériurs. *Comptes rendus des séances de l'Académie des Sciences Paris* 43:219–23.

Lawrence, J., and R. E. Lee. 1955. *Inherit the Wind*. New York: Random House.

Lawrence, P. J. 1972. Edward Hitchcock: The Christian geologist. *Proceedings of the American Philosophical Society* 116:21–34.

Lee, R. 1833. *Memoirs of Baron Cuvier*. London: Longman, Rees, Orme, Brown, Green & Longman.

Leuckart, F. S. 1819. *Zoologische Bruchstücke*. Vol. 1. Helmstädt.

———. 1841. *Zoologische Bruchstücke*. Vol. 2. Stuttgart: F. L. Rieger.

———. 1842. *Zoologische Bruchstücke*. Vol. 3. Freiberg: Gebrüder Groos.

Lewis, G. 1868. *Natural History of Birds: Lectures on Ornithology*. Philadelphia: Bancroft.

Linder, C. R., and L. H. Rieseberg. 2004. Reconstructing patterns of reticulate evolution in plants. *American Journal of Botany* 91:1700–1708.

Linnaeus, C. 1758. *Systema naturæ per regna tria naturæ, secundum classes, ordines, genera, species, cum characteribus, differentiis, synonymis, locis*. Vol. 1. 10th ed. Holmae: Laurentii Salvii.

Longrich, N. R., T. Tokaryk, and D. J. Field. 2011. Mass extinction of birds at the Cretaceous–Paleogene (K–Pg) boundary. *Proceedings of the National Academy of Sciences* 108:15253–57.

Lovejoy, A. O. (1936) 1942. *The Great Chain of Being: A Study of the History of an Idea*. Cambridge, Mass.: Harvard University Press.

Lucretius (Titus Lucretius Carus). 2006. *Lucretius: De rerum natura (On the Nature of Things)*. Translated by W. H. D. Rouse. Revised by M. F. Smith. Loeb Classical Library 181. Cambridge, Mass.: Harvard University Press.

Lull, R. S. 1908. The evolution of the elephant. *American Journal of Science*, 4th ser., 25:169–212.

Lyell, C. 1863. *Geological Evidences of the Antiquity of Man*. London: John Murray.

MacFadden. B. J. 1992. *Fossil Horses: Systematics, Paleobiology, and Evolution of the Family Equidae*. Cambridge: Cambridge University Press.

Macleay, W. S. 1819–1821. *Horae Entomologicae, or Essays on the Annulose Animals*. London: S. Bagster.

Maddison, D. R., and K.-S. Schulz. 2007. The Tree of Life (ToL) Web Project. http://tolweb.org/tree/phylogeny.html.

Maddison, D. R., K.-S. Schulz, and W. P. Maddison. 2007. The Tree of Life Web Project. In "Linnaeus tercentenary: Progress in invertebrate taxonomy," edited by Z.-Q. Zhang and W. A. Shear. Special issue, *Zootaxa* 1668:1–766.

Marsh, O. C. 1879. Polydactyl horses, recent and extinct. *American Journal of Science and Arts* 17:499–505.

Martin, W. C. L. 1841. *A General Introduction to the Natural History of Mammiferous Animals, with a Particular View of the Physical History of Man, and the More Closely Allied Genera of the Order Quadrumana, or Monkeys.* London: Wright.

Mayr, E. 1982. *The Growth of Biological Thought: Diversity, Evolution, and Inheritance.* Cambridge, Mass.: Belknap Press of Harvard University Press.

Mendel, G. 1866. Versuche über Pflanzen-Hybriden. *Verhandlungendes naturforschenden Vereines in Brunn* 4:3–47.

Mikulinskii, S. R., ed. 1972. *Istoriya Biologii s Dryevnyeyshih Vryemyen do Nachala XX Veka* [History of biology from ancient times to the beginning of the twentieth century]. Moscow: Nauka.

Miller, H. 1857. *The Testimony of the Rocks; or, Geology in Its Bearings on the Two Theologies, Natural and Revealed.* Boston: Gould & Lincoln.

Mivart, St. G. 1865. Contributions towards a more complete knowledge of the axial skeleton in the primates. *Proceedings of the Zoological Society of London* 33:545–92.

——. 1867. On the appendicular skeleton of the primates. *Philosophical Transactions of the Royal Society of London* 157:299–429.

——. 1871. *On the Genesis of Species.* New York: Appleton.

Morgan, G. J. 1998. Emile Zuckerkandl, Linus Pauling, and the molecular evolutionary clock, 1959–1965. *Journal of the History of Biology* 31:155–78.

Murphy, W. J., E. Eizirik, S. J. O'Brien, O. Madsen, M. Scally, C. J. Douady, E. Teeling, O. A. Ryder, M. J. Stanhope, W. W. de Jong, and M. S. Springer. 2001. Resolution of the early placental mammal radiation using Bayesian phylogenetics. *Science* 294:2348–51.

Murphy, W. J., D. M. Larkin, A. Everts-van der Wind, G. Bourque, G. Tesler, L. Auvil, J. E. Beever, B. P. Chowdhary, F. Galibert, L. Gatzke, C. Hitte, S. N. Meyers, D. Milan, E. A. Ostrander, G. Pape, H. G. Parker, T. Raudsepp, M. B. Rogatcheva, L. B. Schook, L. C. Skow, M. Welge, J. E. Womack, S. J. O'Brien, P. A. Pevzner, and H. A. Lewin. 2005. Dynamics of mammalian chromosome evolution inferred from multispecies comparative maps. *Science* 309:613–17.

Nelson, G., and N. Platnick. 1981. *Systematics and Biogeography: Cladistics and Vicariance.* New York: Columbia University Press.

Nigro, R. 2000. Piccole storie d'arte: Pantaleone, un mago: Era un umile frate l'autore dello strabiliante mosaico dell'Annunziata a Otranto. *Bell'Italia* 169:90–103.

Norell, M. A. 1992. Taxic origin and temporal diversity: The effect of phylogeny. In *Extinction and Phylogeny*, edited by M. J. Novacek and Q. D. Wheeler, 88–118. New York: Columbia University Press.

Novacek, M. J. 1992. Mammalian phylogeny: Shaking the tree. *Nature* 356:121–25.

Nuttall, G. H. F. 1901. The new biological test for blood in relation to zoological classification. *Proceedings of the Royal Society* 69:150–53.

——. 1904. *Blood Immunity and Blood Relationship: A Demonstration of Certain Blood-Relationships Amongst Animals by Mean of Precipitin Test for Blood.* Cambridge: Cambridge University Press.

Nützel, A., and K. Bandel. 1993. Studies on the side-branch planorbids (Mollusca, Gastropoda) of the Miocene crater lake of Steinheim am Albuch (southern Germany). Special issue, *Scripta Geologica* 2:313–57.

O'Hara, R. J. 1991. Representations of the Natural System in the nineteenth century. *Biology and Philosophy* 6:255–74.

Oldroyd, D. 2001. William Sharp MacLeay. *Encyclopedia of Life Sciences.* http://onlinelibrary.wiley.com/doi/10.1038/npg.els.0.

O'Leary, M. A., J. I. Bloch, J. J. Flynn, T. J. Gaudin, A. Giallombardo, N. P. Giannini, S. L. Goldberg, P. P. Kraatz, Z.-X Luo, J. Meng, X. Ni, M. J. Novacek, F. A. Perini, Z. S. Randall, G. W. Rougier, E. J. Sargis, M. T. Silcox, N. B. Simmons, M. Spaulding, P. M. Velazco, M. Weksler, J. R. Wibke, and A. L. Cirranello. 2013a. The placental mammal ancestor and the post–K-Pg radiation of placentals. *Science* 339: 662–67.

———. 2013b. Response to comment on "The placental mammal ancestor and the post–K-Pg radiation of placentals." *Science* 341:613c.

O'Malley, M. A., ed. 2010. The tree of life. Special issue, *Biology & Philosophy* 25:441–736.

O'Malley, M. A., W. Martin, and J. Dupré. 2010. The tree of life: Introduction to an evolutionary debate. In "The tree of life," edited by M. A. O'Malley. Special issue, *Biology & Philosophy* 25:441–53.

Oppenheimer, J. M. 1987. Haeckel's variations on Darwin. In *Biological Metaphor and Cladistic Classification: An Interdisciplinary Perspective*, edited by H. M. Hoenigswald and L. F. Wiener, 123–35. Philadelphia: University of Pennsylvania Press.

Osborn, H. F. 1893. The rise of the Mammalia in North America. Studies from the Biological Laboratories of Columbia College. *Zoology* 1:1–45.

———. 1905. Origin and history of the horse. Address before the New York Farmers, Metropolitan Club, December 19, Mr. John S. Barnes, presiding.

———. 1910. *The Age of Mammals in Europe, Asia and North America*. New York: Macmillan.

———. 1915. *Men of the Old Stone Age: Their Environment, Life and Art*. New York: Scribner.

———. 1929. *The Titanotheres of Ancient Wyoming, Dakota, and Nebraska*. United States Geological Survey Monograph 55. 2 vols. Washington, D.C.: U.S. Geological Survey.

———. 1933. Aristogenesis, the observed order of biomechanical evolution. *Proceedings of the National Academy of Sciences* 19:699–703.

———. 1934. Aristogenesis, the creative principle in the origin of species. *American Naturalist* 68:193–235.

———. 1936. *Proboscidea: A Monograph of the Discovery, Evolution, Migration and Extinction of the Mastodonts and Elephants of the World*. Vol. 1. New York: American Museum Press.

———. 1942. *Proboscidea: A Monograph of the Discovery, Evolution, Migration and Extinction of the Mastodonts and Elephants of the World*. Vol. 2. New York: American Museum Press.

Osborn, H. F., W. B. Scott, and F. Speir Jr. 1878. *Palaeontological Report of the Princeton Scientific Expedition of 1877*. Contributions from the Museum of Geology and Archaeology of Princeton College, no. 1. New York: Green.

Ovid. 1931. *Fasti*. Translated by J. G. Frazer. Revised by G. P. Goold. Loeb Classical Library 253. Cambridge, Mass.: Harvard University Press.

Owen, R. 1860. *Palaeontology or a Systematic Summary of Extinct Animals and Their Geological Relations*. Edinburgh: Adam and Charles Black.

Packard, A. S. 1901. *Lamarck, the Founder of Evolution: His Life and Work with Translations of His Writings on Organic Evolution*. New York: Longmans, Green.

Pallas, P. S. 1766. *Elenchus zoophytorum sistens generum adumbrationes generaliores et specierum cognitarum succinctas descriptiones cum selectis auctorum synonymis*. The Hague: Petrum van Cleef.

Pennisi, E. 2003. Modernizing the tree of life. *Science* 300:1692–97.

———. 2008. Building the tree of life, genome by genome. *Science* 320:1716–17.

Pietsch, T. W. 2012. *Trees of Life: A Visual History of Evolution*. Baltimore: Johns Hopkins University Press.

Piggin, J. B. 2012. Rereading Isidore. Notes on the History of the Stemma. http://www.piggin.net/stemmahist/isidore.htm.

Placcius, V. 1682. *Justiniani Institutiones juris reconcinnatae*. Frankfurt/Main: Zunner.

Pliny. 1952. *Natural History*. Vol. 9, books 33–35. Translated by H. Rackham. Loeb Classical Library 394. Cambridge, Mass.: Harvard University Press.

——. 2004–2006. *Natural History*. Vols. 1–10, books 1–37. Translated by H. Rackham, W. H. Jones, and D. E. Eichholz. Cambridge, Mass.: Harvard University Press.

Ragan, M. A. 2009. Trees and networks before and after Darwin. *Biology Direct* 4:43. doi:10.1186/1745-6150-4-43.

Rainger, R. 1991. *An Agenda for Antiquity: Henry Fairfield Osborn and Vertebrate Paleontology at the American Museum of Natural History, 1890–1935*. Tuscaloosa: University of Alabama Press.

Rasser, M. W. 2006. 140 Jahre Steinheimer Schnecken-Stammbaum: der älteste fossile Stammbaum aus heutiger Sicht. *Geologica et Palaeontologica* 40:195–99.

Redfield, A. M. 1857. *A General View of the Animal Kingdom*. New York: Kellogg.

——. 1858. *Zoölogical Science, or, Nature in Living Forms, Illustrated by Numerous Plates. Adapted to Elucidate the Chart of the Animal Kingdom*. New York: Kellogg.

Reeves, M. 1999. *Joachim of Fiore and the Prophetic Future*. Guilford, Eng.: Sutton.

Reeves, M., and B. Hirsch-Reich. 1972. *The Figurae of Joachim of Fiore*. Oxford: Oxford University Press.

Regal, B. 2002. *Henry Fairfield Osborn: Race and the Search for the Origins of Man*. Aldershot, Eng.: Ashgate.

Richards, R. J. 2004. If This Be Heresy: Haeckel's Conversion to Darwinism. In *Darwinian Heresies*, edited by A. Lustig, R. J. Richards, and M. Ruse, 101–30. Cambridge: Cambridge University Press.

——. 2008. *The Tragic Sense of Life: Ernst Haeckel and the Struggle over Evolutionary Thought*. Chicago: University of Chicago Press.

Romer, A. S. 1933. *Vertebrate Paleontology*. Chicago: University of Chicago Press.

——. 1971. *Vertebrate Paleontology*. 3rd ed. Chicago: University of Chicago Press.

Romer, A. S., and T. S. Parsons. 1977. *The Vertebrate Body*. 5th ed. Philadelphia: Saunders.

Rosenberg, D., and A. Grafton. 2010. *Cartographies of Time: A History of the Timeline*. New York: Princeton Architectural Press.

Rossini, O. 2008. *Ara Pacis*. Milan: Mondadori Electa.

Rudwick, M. J. S. 1998. *Georges Cuvier, Fossil Bones, and Geological Catastrophes: New Translations and Interpretations of the Primary Texts*. Chicago: University of Chicago Press.

Rütimeyer, L. 1862. Eocaene Säugethiere aus dem Gebiet des Schweizerischen Jura. *Neue Denkschriften der allgemeinen schweizerischen Gesellschaft für die gesammten Naturwissenschaften* 19:1–98.

Safarik, E. A. 2009. *Palazzo Colonna*. Rome: De Luca Editori d'Arte.

Sagan, L. 1967. On the origin of mitosing cells. *Journal of Theoretical Biology* 14:225–74.

Sarich, V. M., and A. C. Wilson. 1967. Immunological time scale for hominid evolution. *Science* 158:1200–1203.

Schedel, H. 1493. *Registrum huius operis libri cronicarum cu[m] figuris et imag[in]ibus ab inicio mu[n]di*. Nuremberg: Koberger.

Schmauch, W. W. 1941. *The Nuremberg Chronicle: Its History and Illustrations. Being the Liber chronicarum of Doctor Hartmann Schedel*. http://www.beloit.edu/nuremberg/index.htm.

Schweigger, A. F. 1820. *Handbuch der Naturgeschichte der skelettlosen ungegliederten Thiere*. Leipzig: Dyk'schen Buchhandlung.

Sclater, P. L. 1864. The mammals of Madagascar. *Quarterly Journal of Science* 1:213–19.

Scopes, J. T., and J. Presley. 1967. *Center of the Storm: Memoirs of John T. Scopes*. New York: Holt, Rinehart and Winston.

Scott, W. B. 1899. The selenodont artiodactyls of the Unita Eocene. *Transactions of the Wagner Free Institute of Science of Philadelphia* 6:ix–xiii, 15–121.

———. 1913. *A History of Land Mammals in the Western Hemisphere*. New York: Macmillan.

Seamon, D. 1998. Goethe, nature, and phenomenology. In *Goethe's Way of Science: A Phenomenology of Nature*, edited by D. Seamon and A. Zajonc, 1–14. Albany: State University of New York Press.

Secord, J. A. 2000. *Victorian Sensation: The Extraordinary Publication, Reception, and Secret Authorship of Vestiges of the Natural History of Creation*. Chicago: University of Chicago Press.

Semes, S. W. 2004. *The Architecture of the Classical Interior*. New York: Norton.

Seneca. 1935. *Moral Essays*. Vol. 3. Translated by J. W. Basore. Loeb Classical Library 310. Cambridge, Mass.: Harvard University Press.

Sibley, C. G., and J. E. Ahlquist. 1984. The phylogeny of the hominoid primates, as indicated by DNA–DNA hybridization. *Journal of Molecular Evolution* 20:2–15.

———. 1987. DNA hybridization evidence of hominid phylogeny: Results from an expanded data set. *Journal of Molecular Evolution* 26:99–121.

———. 1990. *Phylogeny and Classification of the Birds of the World: A Study in Molecular Evolution*. New Haven, Conn.: Yale University Press.

Sibley, C. G., J. E. Ahlquist, and B. L. Monroe Jr. 1988. A classification of the living birds of the world based on DNA–DNA hybridization studies. *The Auk* 105:409–23.

Sibley, C. G., and P. A. Johnsgard. 1959a. An electrophoretic study of egg-white proteins in twenty-three breeds of the domestic fowl. *American Naturalist* 93:107–15.

———. 1959b. Variability in the electrophoretic patterns of avian serum proteins. *Condor* 61:85–95.

Sibley, C. G., and B. L. Monroe Jr. 1990. *Distribution and Taxonomy of Birds of the World*. New Haven, Conn.: Yale University Press.

Simpson, G. G. 1937a. *The Fort Union of the Crazy Mountain Field, Montana, and Its Mammalian Faunas*. United States National Museum, Bulletin, no. 169. Washington, D.C.: Government Printing Office.

———. 1937b. Patterns of phyletic evolution. *Geological Society of America Bulletin* 48:303–14.

———. 1944. *Tempo and Mode in Evolution*. New York: Columbia University Press.

———. 1949. *The Meaning of Evolution*. New Haven, Conn.: Yale University Press.

———. 1961. *Principles of Animal Taxonomy*. New York: Columbia University Press.

———. 1963. The meaning of taxonomic statements. In *Classification and Human Evolution*, edited by S. L. Washburn, 1–31. Chicago: Aldine.

———. 1978. *Concession to the Improbable: An Unconventional Autobiography*. New Haven, Conn.: Yale University Press.

Sneath, P. H. A., and R. R. Sokal. 1973. *Numerical Taxonomy*. San Francisco: Freeman.

Sokal, R. R., and P. H. A. Sneath. 1963. *Principles of Numerical Taxonomy*. San Francisco: Freeman.

Soltis, P. S., and D. E. Soltis. 2009. The role of hybridization in plant speciation. *Annual Review of Plant Biology* 60:561–68.

Springer, M. S., R. W. Meredith, E. C. Teeling, and W. J. Murphy. 2013. Technical comment on "The placental mammal ancestor and the post–K-Pg radiation of placentals." *Science* 341:613b.

Stevens, P. F. 1983. Augustin Augier's "Arbre botanique" (1801), a remarkably early botanical representation of the natural system. *Taxon* 32:203–11.

———. 1994. *The Development of Biological Systematics: Antoine-Laurent de Jussieu, Nature, and the Natural System*. New York: Columbia University Press.

Stoneking, M. 2008. Human origins: The molecular perspective. *EMBO Reports* 9:S46–S50.

Strickland, H. E. 1841. On the true method of discovering the natural system. *Annals and Magazine of Natural History* 6:184–94.

Strong, E. 1858. *History of Bromley in Kent and the Surrounding Neighbourhood Together with an Account of the Colleges, Their Founders, Benefactors etc.* Bromley [London]: Edward Strong.

Taquet, P. 2006. *Georges Cuvier: Naissance d'un génie.* Paris: Odile Jacob.

——. 2009. Cuvier's attitude toward creation and the biblical Flood. In *Geology and Religion: A History of Harmony and Hostility,* edited by M. Kölbl-Ebert, 127–34. Special Publication, no. 310. London: Geological Society.

Tassy, P. 2006. Albert Gaudry et l'émergence de la paléontologie darwinienne au XIXe siècle. *Annales de Paléontologie* 92:41–70.

——. 2011. Trees before and after Darwin. *Journal of Zoological Systematics and Evolutionary Research* 49:89–101.

Tcherikover, A. 1997. *High Romanesque Sculpture in the Duchy of Aquitaine, c. 1090–1140.* Oxford: Clarendon Press.

Tennessee, State of. 1925. House Bill No. 185. An act prohibiting the teaching of Evolution Theory. http://law2.umkc.edu/faculty/projects/ftrials/scopes/tennstat.htm.

Toubert, H. 1970. Le renouveau paléochrétien à Rome au début du XIIe siècle. *Cahiers Archéologiques* 20:99–154.

[Trial Transcript]. 1990 [1925]. *The World's Most Famous Court Trial: Tennessee Evolution Case.* Dayton, Tenn.: Bryan College.

Valadés, D. 1579. *Rhetorica Christiana ad concionandi et orandi usum.* Perugia: P. Petrutium.

van Wyhe, J. 2007. Mind the gap: Did Darwin avoid publishing his theory for many years? *Notes and Records of the Royal Society* 61:177–205.

Velleius Paterculus. 1924. *Compendium of Roman History. Res gestae Divi Augusti.* Translated by F. W. Shipley. Loeb Classical Library 152. Cambridge, Mass.: Harvard University Press.

Vicchi, R. 1999. *The Major Basilicas of Rome: Saint Peter's, San Giovanni, San Paolo, Santa Maria Maggiore.* Florence: Scala.

Vickers, M. 1975. Mantegna and the Ara Pacis. *J. Paul Getty Museum Journal* 2:109–20.

Voss, J. 2010. *Darwin's Pictures: Views of Evolutionary Theory, 1837–1874.* Translated by L. Lantz. New Haven, Conn.: Yale University Press.

Vucinich, A. 1989. *Darwin in Russian Thought.* Berkeley: University of California Press.

Wallace, A. R. 1856. Attempts at a natural arrangement of birds. *Annals and Magazine of Natural History,* 2nd ser, 18:193–216.

——. 1876. *The Geographical Distribution of Animals; with a Study of the Relations of Living and Extinct Faunas as Elucidating the Past changes of the Earth's Surface.* 2 vols. London: Macmillan.

Washburn, S. L., ed. 1963. *Classification and Human Evolution.* Chicago: Aldine.

Wiley, E. O. 1981. *Phylogenetics: The Theory and Practice of Phylogenetic Systematics.* New York: Wiley.

Williams, D. M., and M. C. Ebach. 2008. *Foundations of Systematics and Biogeography.* New York: Springer.

Woese, C. R. 1987. Bacterial evolution. *Microbiology and Molecular Biology Reviews* 51:221–71.

Woese, C. R., and G. E. Fox. 1977. Phylogenetic structure of the prokaryotic domain: The primary kingdoms. *Proceedings of the National Academy of Sciences* 74:5088–90.

Wolpoff, M. H. 1968. "*Telanthropus*" and the single species hypothesis. *American Anthropologist* 70:447–93.

Wolpoff, M. H., J. N. Spuhler, F. H. Smith, J. Radovčić, G. Pope, W. D. Frayer, R. Eckhardt, and G. Clark. 1988. Modern human origins. *Science* 241:772–73.

Wood, B. 2012. A humble pioneer. *Sideways Look* blog, April 10, 2012. http://cashp.gwu.edu/blog/2012/04/10/a-humble-pioneer/.

Zuckerkandl, E., and L. Pauling. 1964. Molecules as documents of evolutionary history [in Russian]. In *Problemy Evolyutsionnoi i Tekhnicheskoi Biokhimii* [Problems of evolutionary and technical biochemistry], edited by V. L. Kretovich, 54–62. Moscow: Nauka.

——. 1965a. Evolutionary divergence and convergence in proteins. In *Evolving Genes and Proteins*, edited by V. Bryson and H. J. Vogel, 97–166. New York: Academic Press.

——. 1965b. Molecules as documents of evolutionary history. *Journal of Theoretical Biology* 8:357–66.

Index